Quantum Mechanics

Third edition

Alastair I. M. Rae

Department of Physics
University of Birmingham
United Kingdom

Institute of Physics Publishing
Bristol and Philadelphia

British Library Cataloguing in Publication Data

A catalogue record for this book is available from the British Library.

ISBN 0 7503 0217 8

Library of Congress Cataloging-in-Publication Data are available

First edition 1980
Second edition 1986
Reprinted 1987
Reprinted with corrections 1990
Reprinted 1991
Third edition 1992

Published by IOP Publishing Ltd, a company wholly owned by the Institute of Physics, London
IOP Publishing Ltd, Techno House, Redcliffe Way, Bristol BS1 6NX, England
US Editorial Office: IOP Publishing Inc., The Public Ledger Building, Suite 1035, Independence Square, Philadelphia, PA 19106

Typeset by P&R Typesetters, Salisbury, Wiltshire
Printed in the UK by Cromwell Press Ltd, Wiltshire

To Angus and Gavin

CONTENTS

PREFACE TO THIRD EDITION

In preparing this edition, I have again gone right through the text identifying points where I thought the clarity could be improved. As a result, numerous minor changes have been made. More major alterations include a discussion of the impressive modern experiments that demonstrate neutron diffraction by macroscopic sized slits in Chapter 1, a revised treatment of Clebsch–Gordan coefficients in Chapter 6 and a fuller discussion of spontaneous emission in Chapter 8. I have also largely rewritten the last chapter on the conceptual problems of quantum mechanics in the light of recent developments in the field as well as of improvements in my understanding of the issues involved and changes in my own viewpoint. This chapter also includes an introduction to the de Broglie–Bohm hidden variable theory and I am grateful to Chris Dewdney for a critical reading of this section.

Alastair I. M. Rae
1992

PREFACE TO SECOND EDITION

I have not introduced any major changes to the structure or content of the book, but I have concentrated on clarifying and extending the discussion at a number of points. Thus the discussion of the application of the uncertainty principle to the Heisenberg microscope has been revised in Chapter 1 and is referred to again in Chapter 4 as one of the examples of the application of the generalized uncertainty principle; I have rewritten much of the section on spin-orbit coupling and the Zeeman effect and I have tried to improve the introduction to degenerate perturbation theory which many students seem to find difficult. The last chapter has been brought up to date in the light of recent experimental and theoretical work on the conceptual basis of the subject and, in response to a number of requests from students, I have provided hints to the solution of the problems at the ends of the chapters.

I should like to thank everyone who drew my attention to errors or suggested improvements. I believe nearly every one of these suggestions has been incorporated in one way or another into this new edition.

Alastair I. M. Rae
1985

PREFACE TO FIRST EDITION

Over the years the emphasis of undergraduate physics courses has moved away from the study of classical macroscopic phenomena towards the discussion of the microscopic properties of atomic and subatomic systems. As a result, students now have to study quantum mechanics at an earlier stage in their course without the benefit of a detailed knowledge of much of classical physics and, in particular, with little or no acquaintance with the formal aspects of classical mechanics. This book has been written with the needs of such students in mind. It is based on a course of about thirty lectures given to physics students at the University of Birmingham towards the beginning of their second year—although, perhaps inevitably, the coverage of the book is a little greater than I was able to achieve in the lecture course. I have tried to develop the subject in a reasonably rigorous way, covering the topics needed for further study in atomic, nuclear, and solid state physics, but relying only on the physical and mathematical concepts usually taught in the first year of an undergraduate course. On the other hand, by the end of their first undergraduate year most students have heard about the basic ideas of atomic physics, including the experimental evidence pointing to the need for a quantum theory, so I have confined my treatment of these topics to a brief introductory chapter.

While discussing those aspects of quantum mechanics required for further study, I have laid considerable emphasis on the understanding of the basic ideas and concepts behind the subject, culminating in the last chapter which contains an introduction to quantum measurement theory. Recent research, particularly the theoretical and experimental work inspired by Bell's theorem, has greatly clarified many of the conceptual problems in this area. However, most of the existing literature is at a research level and concentrates more on a rigorous presentation of results to other workers in the field than on making them accessible to a wider audience. I have found that many physics undergraduates are particularly interested in this aspect of the subject and there is therefore a need for a treatment suitable for this level. The last chapter of this book is an attempt to meet this need.

I should like to acknowledge the help I have received from my friends and colleagues while writing this book. I am particularly grateful to Robert Whitworth, who read an early draft of the complete book, and to Goronwy Jones and George Morrison, who read parts of it. They all offered many valuable and penetrating criticisms, most of which have been incorporated in this final version. I should also like to thank Ann Aylott who typed the manuscript and was always patient and helpful throughout many changes and revisions, as well as Martin Dove who assisted with the proofreading. Naturally, none of this help in any way lessens my responsibility for whatever errors and omissions remain.

Alastair I. M. Rae
1980

ONE

INTRODUCTION

Quantum mechanics was developed as a response to the inability of the classical theories of mechanics and electromagnetism to provide a satisfactory explanation of some of the properties of electromagnetic radiation and of atomic structure. As a result a theory has emerged, whose basic principles can be used to explain not only the structure and properties of atoms, including the way they interact with each other in molecules and solids, but also those of nuclei and of 'elementary' particles such as the proton and neutron. Although there are still many features of the physics of such systems that are not fully understood, there are presently no indications that the fundamental ideas of quantum mechanics are incorrect. In order to achieve this success, quantum mechanics has been built on a foundation that contains a number of concepts that are fundamentally very different from those of classical physics and which have completely altered our view of the way the natural universe operates. This book will attempt to elucidate and discuss the conceptual basis of the subject as well as explaining its success in describing the behaviour of atomic and subatomic systems.

Quantum mechanics is often thought to be a difficult subject, not only in its conceptual foundation, but also in the complexity of its mathematics. However, although a rather abstract formulation is required for a proper treatment of the subject, much of the apparent complication arises in the course of the solution of essentially simple mathematical equations applied to particular physical situations. We shall discuss a number of such applications in this book, because it is important to appreciate the success of quantum mechanics in explaining the results of real physical measurements.

However, the reader should try not to allow the ensuing algebraic complication to hide the essential simplicity of the basic ideas.

In this first chapter we shall discuss some of the key experiments that illustrate the failure of classical physics. However, although the experiments described were performed in the first quarter of this century and played an important role in the development of the subject, we shall not be giving a historically based account. Neither will our account be a complete description of the early experimental work; for example, we shall not describe the experiments on the properties of thermal radiation and the heat capacity of solids that provided early indications of the need for the quantization of the energy of electromagnetic radiation and of mechanical systems. The topics to be discussed have been chosen as those that point most clearly towards the basic ideas needed in the further development of the subject. As so often happens in physics, the way in which the theory actually developed was by a process of trial and error, often relying on flashes of inspiration, rather than the more logical approach suggested by hindsight.

1.1 THE PHOTOELECTRIC EFFECT

When light strikes a clean metal surface in a vacuum, it causes electrons to be emitted with a range of energies. For light of a given frequency v the maximum electron energy E_x is found to be equal to the difference between two terms, one of which is proportional to the frequency of the incident light with a constant of proportionality h that is the same whatever the metal used, while the other is independent of frequency but varies from metal to metal. Moreover, neither term depends on the intensity of the incident light which affects only the rate of electron emission. Thus

$$E_x = hv - \phi \tag{1.1}$$

It is very difficult, if not impossible, to explain this result on the basis of the classical theory of light as an electromagnetic wave. This is because the energy contained in such a wave would arrive at the metal at a uniform rate and there is no apparent reason why this energy should be divided up in such a way that the maximum electron energy is proportional to the frequency and independent of the intensity of the light. A very important feature of the photoelectric effect is the dependence of the rate of electron emission on the light intensity. Although the average emission rate is proportional to the intensity, individual electrons are emitted at random, and when experiments are performed using very weak light, electrons are sometimes emitted well before sufficient electromagnetic energy should have arrived at the metal.

Such considerations led Einstein to postulate that the classical electromagnetic theory does not provide a complete explanation of the properties of light, and that we must also assume that the energy in an electromagnetic wave is 'quantized' in the form of small packets, known as *photons*, each of which carries an amount of energy equal to $h\nu$. Given this postulate, we can see that when light is incident on a metal, the maximum energy an electron can gain is that carried by one of the photons. Part of this energy will be given up by the electron as it escapes from the metal surface—so accounting for the quantity ϕ in (1.1), which is accordingly known as the *work function*—but the rest will be converted into the kinetic energy of the freed electron, in agreement with the experimental results summarized in Eq. (1.1). The photon postulate also explains the emission of photoelectrons at random times. Thus, although the average rate of photon arrival is proportional to the light intensity, individual photons arrive at random and, as each carries with it a quantum of energy, there will be occasions when an electron is emitted well before this would be classically expected.

The constant h connecting the energy of a photon with the frequency of the electromagnetic wave is known as *Planck's constant*, because it was originally postulated by Planck in order to explain some of the properties of thermal radiation. It is a fundamental constant that frequently occurs in the equations of quantum mechanics. We shall find it convenient to change this notation slightly and define another constant $\hbar$ as being equal to h divided by 2π; also, when referring to waves, we shall normally use the angular frequency ω ($= 2\pi\nu$), in preference to the frequency ν. Using this notation, the photon energy E can be expressed as

$$E = \hbar\omega \qquad (1.2)$$

Throughout this book we shall write our equations in terms of $\hbar$ and avoid ever again referring to h. We note that $\hbar$ has the dimensions of action (energy × time) and its currently accepted value is $(1.054\,589 \pm 0.000\,006) \times 10^{-34}$ J s.

1.2 THE COMPTON EFFECT

The existence of photons is also demonstrated by experiments first carried out by A. H. Compton that involve the scattering of X-rays by electrons. To understand these we must make the further postulate that a photon, as well as carrying a quantum of energy, also has a definite momentum and can therefore be treated in many ways just like a classical particle. An expression for the photon momentum is suggested by the classical theory of radiation pressure: it is known that if energy is transported by an electromagnetic wave at a rate W per unit area per second, then the wave exerts a pressure of magnitude W/c (where c is the velocity of light), whose direction is parallel to that of the wave vector $\mathbf{k}$ of the wave; if we now treat the wave as composed of photons of energy $\hbar\omega$ it follows that the photon momentum $\mathbf{p}$ should have a magnitude $\hbar\omega/c = \hbar k$ and that its direction should be parallel

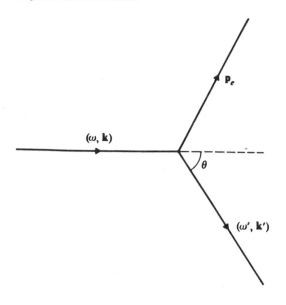

Figure 1.1 In Compton scattering an X-ray photon of angular frequency ω and wave vector **k** collides with an electron initially at rest. After the collision the photon frequency and wave vector are changed to ω' and **k'** respectively and the electron recoils with momentum $\mathbf{p}_e$.

to **k**. Thus

$$\mathbf{p} = \hbar\mathbf{k} \tag{1.3}$$

We now consider a collision between such a photon and an electron of mass m that is initially at rest. After the collision we assume that the frequency and wave vector of the photon are changed to ω' and **k'** and that the electron moves off with momentum $\mathbf{p}_e$ as shown in Fig. 1.1. Assuming that energy and momentum are conserved we have

$$\hbar\omega - \hbar\omega' = p_e^2/2m \tag{1.4}$$

$$\hbar\mathbf{k} - \hbar\mathbf{k}' = \mathbf{p}_e \tag{1.5}$$

Squaring (1.5) and substituting into (1.4) we get

$$\hbar(\omega - \omega') = \frac{\hbar^2}{2m}(\mathbf{k} - \mathbf{k}')^2$$

$$= \frac{\hbar^2}{2m}[(k - k')^2 + 2kk'(1 - \cos\theta)] \tag{1.6}$$

where θ is the angle between **k** and **k'** (cf. Fig. 1.1). Now the change in the magnitude of the wave vector $(k - k')$ always turns out to be very much smaller than either k or k' so we can neglect the first term in square brackets on the right-hand side of (1.6). Remembering that $\omega = ck$ and $\omega' = ck'$ we then get

$$\frac{1}{\omega'} - \frac{1}{\omega} = \frac{\hbar}{mc^2}(1 - \cos\theta)$$

that is
$$\lambda' - \lambda = \frac{2\pi\hbar}{mc}(1 - \cos\theta) \qquad (1.7)$$

where λ and λ' are the X-ray wavelengths before and after the collision, respectively. It turns out that if we allow for relativistic effects when carrying out the above calculation, we obtain the same result as (1.7) without having to make any approximations.

Experimental studies of the scattering of X-rays by electrons in solids produce results in good general agreement with the above predictions. In particular, if the intensity of the radiation scattered through a given angle is measured as a function of the wavelength of the scattered X-rays, a peak is observed whose maximum lies just at the point predicted by (1.7). In fact such a peak has a small, but finite, width implying that some of the photons have been scattered in a manner slightly different from that described above, but this can be explained by taking into account the fact that the electrons in a solid are not necessarily at rest, but generally have a finite momentum before the collision. Compton scattering can therefore be used as a tool to measure the electron momentum, and we shall discuss this in more detail in Chapter 4.

Both the photoelectric effect and the Compton effect are connected with the interactions between electromagnetic radiation and electrons, and both provide conclusive evidence for the photon nature of electromagnetic waves. However, we might ask why there are two effects and why the X-ray photon is scattered by the electron with a change of wavelength, while the optical photon transfers all its energy to the photoelectron. The principal reason is that in the X-ray case the photon energy is much larger than the binding energy between the electron and the solid; the electron is therefore knocked cleanly out of the solid in the collision and we can treat the problem by considering energy and momentum conservation. In the photoelectric effect, on the other hand, the photon energy is only a little larger than the binding energy and, although the details of this process are rather complex, it turns out that the momentum is shared between the electron and the atoms in the metal and that the whole of the photon energy is used to free the electron and give it kinetic energy. However, none of these detailed considerations affects the conclusion that in both cases the incident electromagnetic radiation exhibits properties consistent with it being composed of photons whose energy and momentum are given by the expressions (1.2) and (1.3).

1.3 LINE SPECTRA AND ATOMIC STRUCTURE

When an electric discharge is passed through a gas, light is emitted which, when examined spectroscopically, is typically found to consist of a series of lines, each of which has a sharply defined frequency. A particularly simple example of such a line spectrum is that of hydrogen, in which case the observed frequencies are given by the formula

$$\omega_{mn} = 2\pi R_0 c \left(\frac{1}{n^2} - \frac{1}{m^2} \right) \qquad (1.8)$$

where n and m are integers, c is the speed of light and R_0 is a constant known as the *Rydberg constant* (after J. R. Rydberg who first showed that the experimental results fitted this formula) whose currently accepted value is $(1.096\,755\,9 \pm 0.000\,000\,1) \times 10^7\,\mathrm{m}^{-1}$.

Following our earlier discussion, we can assume that the light emitted from the atom consists of photons whose energies are $\hbar\omega_{mn}$. It follows from this and the conservation of energy, that the energy of the atom emitting the photon must have been changed by the same amount, and the obvious conclusion to draw is that the energy of the hydrogen atom is itself quantized so that it can adopt only one of the values E_n where

$$E_n = -\frac{2\pi R_0 \hbar c}{n^2} \qquad (1.9)$$

the negative sign corresponding to the negative binding energy of the electron in the atom. Similar constraints govern the values of the energies of atoms other than hydrogen although these cannot usually be expressed in such a simple form. We refer to allowed energies such as E_n as *energy levels*. Further confirmation of the existence of energy levels is obtained from the ionization energies and absorption spectra of atoms, which both display features consistent with the energy of an atom being quantized in this way. It will be one of the main aims of this book to develop a theory of quantum mechanics that will successfully explain the existence of energy levels and provide a theoretical procedure for calculating their values.

One feature of the structure of atoms that can be at least partly explained on the basis of energy quantization is the simple fact that atoms exist at all! According to classical electromagnetic theory, an accelerated charge always loses energy in the form of radiation, so a negative electron in motion about a positive nucleus should radiate, lose energy, and quickly coalesce with the nucleus. The fact that the radiation is quantized should not affect this argument, but if the energy of the atom is quantized, there will be a minimum energy level (that with $n = 1$ in the case of hydrogen) below which the atom cannot go and in which it will remain indefinitely. This also explains why all atoms of the same species behave in the same way. As we shall see later, all hydrogen atoms in the lowest energy state have the same properties. This is in contrast to a classical system, such as a planet orbiting a star, in which case an infinite number of possible orbits with very different properties can exist for a given value of the energy of the system.

1.4 DE BROGLIE WAVES

Following on from the fact that the photons associated with electromagnetic waves behave like particles, L. de Broglie suggested that particles such as electrons might also have wave properties. He further proposed that the frequencies and wave vectors of these 'matter waves' would be related to the energy and momentum of the associated particle in the same way as in the photon case. That is

$$\left.\begin{array}{r} E = \hbar\omega \\ \mathbf{p} = \hbar\mathbf{k} \end{array}\right\} \tag{1.10}$$

In the case of matter waves, Eqs (1.10) are referred to as the de Broglie relations. We shall develop this idea in subsequent chapters when we shall find that it can be used to account for atomic energy levels. In the meantime we shall describe a simple experiment that provides direct confirmation of the existence of matter waves.

The property possessed by a wave that distinguishes it from any other physical phenomenon is its ability to form interference and diffraction patterns: when different parts of a wave are recombined after travelling different distances they reinforce each other or cancel out depending on whether the two path lengths differ by an even or an odd number of wavelengths. Such phenomena are readily demonstrated in the laboratory by passing light through a diffraction grating for example. Unfortunately, if the wavelength of the waves associated with even very low energy electrons (say around 1 eV) is calculated using the de Broglie relations (1.10) a value of around 10^{-9} m is obtained which is much smaller than that of visible light and much too small to form a detectable diffraction pattern when passed through a conventional grating. However, the diffraction of very short wave electromagnetic radiation (X-rays) can be demonstrated using crystals where the separation of neighbouring planes of atoms is of the order of 10^{-9} m. As shown in Fig. 1.2 diffraction occurs when the incident and scattered X-rays make the same angle θ with the atomic planes and when the wavelength λ of the de Broglie wave is related to the interplanar separation d and to θ by the Bragg equation.

$$n\lambda = 2d \sin \theta \tag{1.11}$$

The first experiment to demonstrate the diffraction of electron waves by crystals was carried out by C. Davisson and L. H. Germer who studied the scattering of beams of electrons by single crystals of nickel. Although the low-energy beams could not penetrate far enough below the crystal surface to exhibit Bragg diffraction as described above, they were diffracted in a similar manner by the two-dimensional atomic lattice in the surface layer, and the diffraction of matter waves was clearly demonstrated. Moreover,

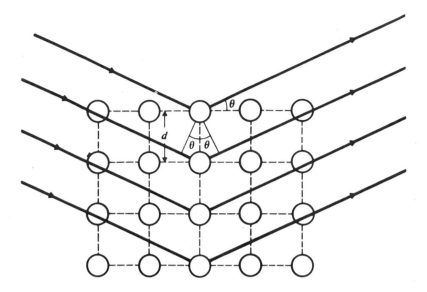

Figure 1.2 Waves incident at an angle θ to planes of atoms in a crystal are diffracted so that the path difference between the de Broglie waves scattered by successive planes is a whole number of wavelengths.

measurements on the diffraction pattern allowed the wavelength of the matter waves to be calculated and this was found to agree with the value predicted by the de Broglie relations (1.10).

Nowadays the wave properties of electron beams are commonly observed experimentally and electron microscopes, for example, are often used to display the diffraction patterns of the objects under observation. Moreover, not only electrons behave in this way; neutrons of the appropriate energy can also be diffracted by crystals, this technique being commonly used to investigate structural and other properties of solids. In recent years, neutron beams have been produced with such low energy that their de Broglie wavelength is as large as 2.0 nm. When these are passed through a double slit whose separation is of the order of 0.1 mm, the resulting diffraction maxima are separated by about 10^{-3} degrees, which corresponds to about 0.1 mm at a distance of 5 m beyond the slits, where the detailed diffraction pattern can be resolved. Figure 1.3 gives the details of such an experiment and the results obtained; we see that the number of neutrons recorded at different angles is in excellent agreement with the intensity of the diffraction pattern, calculated on the assumption that the neutron beam can be represented by a de Broglie wave.

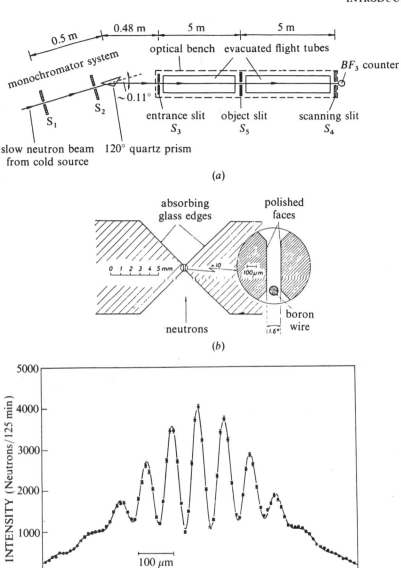

Figure 1.3 In recent years, it has been possible to produce neutron beams with de Broglie wavelengths around 2 nm which can be detectably diffracted by double slits of separation about 0.1 mm. A typical experimental arrangement is shown in (a) and the slit arrangement is illustrated in (b). The number of neutrons recorded along a line perpendicular to the diffracted beam 5 m beyond the slits is shown in (c), along with the intensity calculated from diffraction theory, assuming a wave model for the neutron beam. The agreement is clearly excellent. (Reproduced by permission from A. Zeilinger, R. Gähler, C. G. Schull, W. Treimer and W. Mampe, *Reviews of Modern Physics* **60** 1067–73 (1988).)

1.5 WAVE-PARTICLE DUALITY

Although we have just described the experimental evidence for the wave nature of electrons and similar bodies, it must not be thought that this. description is complete or that these are any-the-less particles. Although in a diffraction experiment wave properties are manifested during the diffraction process and the intensity of the wave determines the average number of particles scattered through various angles, when the diffracted electrons are detected they are always found to behave just like point particles with the expected mass and charge and to have a particular energy. Conversely, although we need to postulate photons in order to explain phenomena such as the photoelectric and Compton effects, phenomena such as the diffraction of light by a grating or of X-rays by a crystal can be explained only if electromagnetic radiation has wave properties.

In many circumstances it is perfectly clear which model should be used in a particular physical situation. Thus, although electrons with a kinetic energy of 100 eV (1.6×10^{-17} J) have a de Broglie wavelength of about 10^{-10} m and are therefore diffracted by crystals according to the wave model, if their energy is very much higher (say 100 MeV) the wavelength is then so short ($c.\ 10^{-14}$ m) that diffraction effects are not normally observed and such electrons nearly always behave like classical particles. A similar argument shows why the wave properties of everyday macroscopic particles are not apparent: even a small grain of sand of mass about 10^{-6} g moving at a speed of 10^{-3} m s^{-1} has a de Broglie wavelength of the order of 10^{-21} m and its wave properties are therefore quite undetectable; clearly this is even more true for heavier or faster moving objects. There are some experimental situations, however, which cannot be understood without involving both the wave and the particle properties.

If we examine the neutron diffraction experiment illustrated in Fig. 1.3, we see how it illustrates this wave-particle duality. The neutron beam behaves like a wave when it is passing through the slits and forming an interference pattern, but when the neutrons are detected, they behave like a set of individual particles with the usual mass, zero electric charge etc. We never detect half a neutron! Moreover, the typical neutron beams used in such experiments are so weak that no more than one neutron is in the apparatus at any one time and we therefore cannot explain the interference pattern on the basis of any model involving interactions between different neutrons.

Suppose we now change the above experiment so as to place detectors behind each slit instead of a large distance away; these will detect individual neutrons passing through one or other of the slits—but never both at once—and the obvious conclusion is that the same thing happened in the interference experiment. But we have just seen that the interference pattern is formed by a wave passing through both slits and this can be confirmed by arranging a system of shutters so that only one or other of the two slits, but never both, are open at any one time, in which case it is impossible to form an interference pattern. Both slits are necessary to form the interference pattern, so if the neutrons always pass through one slit or the other then the behaviour of a given neutron must somehow be affected by the slit it did not pass through!

An alternative view, which is now the orthodox interpretation of quantum mechanics, is to say that the model we use to describe quantum phenomena is not just a property of the quantum objects (the neutrons in this case) but also depends on the arrangement of the whole apparatus. Thus, if we perform a diffraction experiment, the neutrons are waves when they pass through the slits, but are particles when they are detected. But if the experimental apparatus includes detectors right behind the slits, the neutrons behave like particles at this point. This dual description is possible because no interference pattern is created in the latter case. Moreover, it turns out that this happens no matter how subtle an experiment we design to detect which slit the neutron passes through: if it is successful, the phase relation between the waves passing through the slits is destroyed and the interference pattern disappears. We can therefore look on the particle and wave models as complementary rather than contradictory properties. Which one is manifest in a particular experimental situation depends on the arrangement of the whole apparatus, including the slits and the detectors, and we cannot conclude that because we detect particles when we place detectors behind the slits, the neutrons will still have these properties when we put them somewhere else.

It should be noted that, although we have just discussed neutron diffraction, the argument would have been unchanged if we had considered light waves and photons or any other particles with their associated waves. In fact the idea of complementarity is even more general than this and we shall find many cases in our discussion of quantum mechanics where the measurement of one property of a physical system renders another unobservable; an example of this will be described in the next paragraph when we discuss the limitations on the simultaneous measurement of the position and momentum of a particle. Many of the apparent paradoxes and contradictions that arise can be resolved, provided we concentrate on those aspects of a physical system that can be directly observed and beware of drawing conclusions about properties that cannot. However, there are still significant conceptual problems in this area which remain the subject of active research, and we shall discuss these in some detail in Chapter 11.

The Uncertainty Principle

In this paragraph we consider the limits that wave-particle duality places on the simultaneous measurement of the position and momentum of a particle.

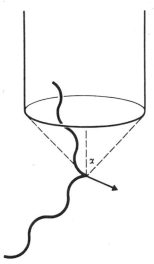

Figure 1.4 A measurement of the position of a particle by a microscope causes a corresponding uncertainty in the particle momentum as it recoils after interaction with the illuminating radiation.

Suppose we try to measure the position of a particle by illuminating it with radiation of wavelength λ and using a microscope of angular aperture α, as shown in Fig. 1.4. The fact that the radiation has wave properties means that an image of the electron will be observed in the microscope, but the size of this image will be governed by the resolving power of the microscope. The position of the electron is therefore uncertain by an amount Δx which is given by standard optical theory as

$$\Delta x \simeq \frac{\lambda}{\sin \alpha} \tag{1.12}$$

On the other hand, the fact that the radiation is composed of photons means that each time the particle is struck by a photon it recoils, as in Compton scattering. The momentum of the recoil could of course be calculated if we knew the initial and final momenta of the photon, but as we do not know through which points on the lens the photons entered the microscope, the x component of the particle momentum is subject to an error Δp_x where

$$\begin{aligned} \Delta p_x &\simeq p \sin \alpha \\ &= 2\pi\hbar \sin \alpha / \lambda \end{aligned} \tag{1.13}$$

Combining (1.12) and (1.13) we get

$$\Delta x \, \Delta p_x \simeq 2\pi\hbar \tag{1.14}$$

It follows that if we try to improve the accuracy of the position measurement by using radiation with a smaller wavelength, we shall increase the error on the momentum measurement and vice versa. It should be noted that in this example the measurement is performed using a light beam containing many photons—otherwise we could not have built up an image in the microscope. Moreover, we have implicitly assumed that all these photons arrived within a short space of time so that we could neglect additional uncertainties in position due to the electron's recoiling motion. This is just one example of an experiment designed to measure the position and momentum of a particle, but it turns out that any other experiment with this aim is subject to constraints similar to (1.14). We shall see in Chapter 4 that it is a direct consequence of the fundamental principles of quantum mechanics that in every case the errors in the position and momentum components are related by

$$\Delta x \, \Delta p_x \geqslant \tfrac{1}{2}\hbar \qquad (1.15)$$

This relation is known as the Heisenberg uncertainty principle. According to quantum mechanics it is a fundamental property of nature that any attempt to make simultaneous measurements of position and momentum are subject to this limitation.

1.6 THE REST OF THIS BOOK

In the next two chapters we discuss the nature and properties of matter waves in more detail and show how to obtain a wave equation whose solutions determine the energy levels of bound systems. We shall do this by considering one-dimensional waves in Chapter 2 where we shall obtain qualitative agreement with experiment, while in the following chapter we shall extend our treatment to three-dimensional systems and obtain excellent quantitative agreement between the theoretical results and experimental values of the energy levels of the hydrogen atom. At the same time we shall find that this treatment is incomplete and leaves many important questions unanswered. Accordingly, in Chapter 4 we shall set up a more formal version of quantum mechanics within which the earlier results are included, but which can be applied more generally; this will prove to be a rather abstract process and prior familiarity with the results discussed in the earlier chapters will be a great advantage in understanding it. Having set up the general theory, it is then developed in subsequent chapters and discussed along with its applications to a number of problems such as the quantum theory of angular momentum and the special properties of systems containing a number of identical particles, while the last chapter contains a detailed discussion of some of the conceptual problems of quantum mechanics. Chapters 7 to 11 are largely, although not entirely, self-contained and can be read in a different order if desired.

Finally we should point out that photons, which have been referred to quite extensively in this chapter, will hardly be mentioned again except in passing. This is primarily because a discussion of the quantization of the electromagnetic field, going beyond the simple treatment in this chapter, requires a degree of mathematical sophistication which is unsuitable for a book at this level. However, it turns out that, despite the importance of the photon in suggesting the idea of quantized waves, many physical phenomena can be understood by considering the mechanical system to be quantized and treating the electromagnetic fields semi-classically. Indeed it has been pointed out that even many of the observed features of the photoelectric and Compton effects can be explained in this way—although only if we are prepared to accept energy conservation as being a statistical property and not necessarily true for an individual quantum system. Nevertheless, there are a number of important phenomena, particularly in high-energy physics, that clearly establish the quantum properties of electromagnetic waves, and field quantization is an essential tool in considering such topics. However, most of the principles of quantum mechanics can be discussed in the context of the quantization of mechanical systems, and this will be the aim of the present book.

PROBLEMS

1.1 The maximum energy of photoelectrons emitted from potassium is 2.1 eV when illuminated by light of wavelength 3×10^{-7} m and 0.5 eV when the light wavelength is 5×10^{-7} m. Use these results to obtain values for Planck's constant and the minimum energy needed to free an electron from potassium.

1.2 If the energy flux associated with a light beam of wavelength 3×10^{-7} m is 10 W m^{-2}, estimate how long it would take, classically, for sufficient energy to arrive at a potassium atom of radius 2×10^{-10} m in order that an electron be ejected. What would be the average emission rate of photoelectrons if such light fell on a piece of potassium 10^{-3} m^2 in area? Would you expect your answer to the latter question to be significantly affected by quantum mechanical considerations?

1.3 An X-ray photon of wavelength 1.0×10^{-12} m is incident on a stationary electron. Calculate the wavelength of the scattered photon if it is detected at an angle of (i) 60°, (ii) 90°, and (iii) 120° to the incident radiation.

1.4 Electrons of energy 100 eV are incident on a crystal containing planes of atoms whose separation is 5×10^{-10} m. For what values of the Bragg angle will diffraction from these planes occur? What happens if the electrons are replaced by neutrons of the same energy?

1.5 A beam of neutrons with known momentum is diffracted by a single slit in a geometrical arrangement similar to that shown for the double slit in Fig. 1.3. Show that an approximate value of the component of momentum of the neutrons in a direction perpendicular to both the slit and the incident beam can be derived from the single-slit diffraction pattern. Show that the uncertainty in this momentum is related to the uncertainty in the position of the neutron passing through the slit in a manner consistent with the Heisenberg uncertainty principle. (This example is discussed in more detail in Chapter 4.)

THE ONE-DIMENSIONAL SCHRÖDINGER EQUATIONS

In the previous chapter we have seen that electrons and other subatomic particles sometimes exhibit properties similar to those commonly associated with classical waves so that, for example, electrons of the appropriate energy are diffracted by crystals in a manner similar to that originally observed in the case of X-rays. Moreover, the energy and momentum of a free particle can be expressed in terms of the angular frequency and wave vector of the associated plane wave by the de Broglie relations (1.10).

We are going to develop these ideas to see how the wave properties of the electrons bound within atoms can account for atomic properties such as line spectra. Clearly atoms are three-dimensional objects, so we shall eventually have to consider three-dimensional waves. However, this involves somewhat complex analysis, so in this chapter we shall begin by studying the properties of electron waves in one dimension.

In one dimension the wave vector and momentum of a particle can be treated as scalars so the de Broglie relations can be written as

$$E = \hbar\omega \qquad p = \hbar k \qquad (2.1)$$

We shall use these and the properties of classical waves to set up a wave equation, known as the *Schrödinger wave equation*, appropriate to these 'matter waves', and when we solve this equation for the case of particles that are not free but move in a potential well, we shall find that solutions are only possible for particular discrete values of the total energy. We shall apply this theory to a number of examples and compare the resulting energy levels with experimental results.

15

2.1 THE TIME-DEPENDENT SCHRÖDINGER EQUATION

Consider a classical plane wave moving along the x axis. Its displacement at the point x at time t is given by the real part of the complex quantity A where

$$A(x, t) = A_0 \exp[i(kx - \omega t)] \qquad (2.2)$$

(In the case of electromagnetic waves, for example, the real part of A equals the magnitude of the electric field vector.) The above expression is the solution to a *wave equation* and the form of wave equation applicable to many classical waves is

$$\frac{\partial^2 A}{\partial x^2} = \frac{1}{c^2} \frac{\partial^2 A}{\partial t^2} \qquad (2.3)$$

where c is a real constant equal to the wave velocity. If we substitute the right-hand side of (2.2) into (2.3), we see that the former is a solution to the latter if

$$-k^2 = -\omega^2/c^2$$

that is
$$\omega = ck \qquad (2.4)$$

We can see immediately that the equation governing matter waves cannot have the form (2.3), because (2.4) combined with the de Broglie relations (2.1) gives the linear relation

$$E = cp \qquad (2.5)$$

whereas for non-relativistic free particles the energy and momentum are known to obey the classical relation

$$E = p^2/2m \qquad (2.6)$$

In the case of matter waves, therefore, we must look for a wave equation of a different kind from (2.3), but, because we know that plane waves are associated with free particles, the expression (2.2) must also be a solution to this new equation.

If the equations (2.1) and (2.6) are to be satisfied simultaneously, it is necessary that the frequency of the wave be proportional to the square of the wave vector, rather than directly proportional as in (2.4). This indicates that a suitable wave equation might involve differentiating twice with respect to x, as in (2.3), but only once with respect to t. Consider, therefore, the equation

$$\frac{\partial^2 \Psi}{\partial x^2} = \alpha \frac{\partial \Psi}{\partial t} \qquad (2.7)$$

where α is a constant and $\Psi(x, t)$ is a quantity known as the *wave function* whose significance will be discussed shortly. If we now substitute a plane wave of the form (2.2) for Ψ we find that this is a solution to (2.7) if

$$-k^2 = -i\alpha\omega$$

We are therefore able to satisfy (2.1) and (2.6) by defining α such that

$$\alpha = -2mi/\hbar$$

Substituting into (2.7) and rearranging slightly we obtain the wave equation for the matter waves associated with free particles as

$$i\hbar\frac{\partial\Psi}{\partial t} = -\frac{\hbar^2}{2m}\frac{\partial^2\Psi}{\partial x^2} \tag{2.8}$$

We can verify that this equation meets all the above requirements by substituting the plane wave form of Ψ and using the de Broglie relations (2.1) when we get

$$E\Psi = (p^2/2m)\Psi \tag{2.9}$$

as expected. However, so far we have only found an equation which produces the correct results for a free particle, whereas we are looking for a more general theory to include the case of a particle moving under the influence of a potential, $V(x, t)$. The total energy E in this case is equal to the sum of the kinetic and potential energies which suggests a possible generalization of (2.9) to

$$E\Psi = (p^2/2m + V)\Psi$$

which in turn suggests that the wave equation (2.9) could be similarly generalized to give

$$i\hbar\frac{\partial\Psi}{\partial t} = -\frac{\hbar^2}{2m}\frac{\partial^2\Psi}{\partial x^2} + V\Psi \tag{2.10}$$

Equation (2.10) is known as the *one-dimensional time-dependent Schrödinger equation*; its further generalization to three-dimensional systems is quite straightforward and will be discussed in the following chapter. We shall shortly obtain solutions to this equation for various forms of the potential $V(x, t)$, but in the meantime we shall pause to consider the validity of the arguments used to obtain (2.10).

It is important to note that in no way do these arguments constitute a rigorous derivation of a result from more basic premises: we started with a limited amount of experimental knowledge concerning the properties of free particles and their associated plane waves, and we ended up with an equation for the wave function associated with a particle moving under the influence of a general potential! Such a process whereby we proceed from a particular example to a more general law is known as *induction*, in contrast with *deduction* whereby a particular result is derived from a more general premise.

Induction is very important in science, and is an essential part of the process of the development of new theories, but it cannot by itself establish the truth, or otherwise, of the general laws so obtained. These remain inspired guesses until physical properties have been deduced from them and found to be in agreement with the results of experimental measurement. Of course, if only one case of disagreement were found, the theory would be falsified and we should need to look for a more general law whose predictions would then have to agree with experiment in this new area, as well as in the other cases where the earlier theory was successful. The Schrödinger equation, and the more general formulation of quantum mechanics to be discussed in Chapter 4, have been set up as a result of the failure of classical physics to predict correctly the results of experiments on microscopic systems; they must be verified by testing their predictions of the properties of systems where classical mechanics has failed and also where it has succeeded. Much of the rest of this book will consist of a discussion of such predictions and we shall find that the theory is successful in every case; in fact the whole of atomic physics, solid state physics, and chemistry obey the principles of quantum mechanics, and the same is true of nuclear and particle physics, although an understanding of very high-energy phenomena requires an extension of the theory to include relativistic effects and this is outside the scope of this book.

The Wave Function

We now discuss the physical significance of the wave function, $\Psi(x, t)$, which was introduced in Eq. (2.7). We first note that, unlike the classical wave displacement, the wave function is essentially a complex quantity; whereas the complex form of the classical wave is used for convenience, physical significance being confined to its real part which is itself a solution to the classical wave equation, neither the real nor the imaginary part of the wave function, but only the full complex expression, is a solution to the Schrödinger equation. It follows that the wave function cannot itself be identified with a single physical property of the system. However, it has an indirect significance which we shall now discuss—again using an inductive argument.

When we discussed diffraction in Chapter 1, we saw that, although the behaviour of the individual particles is random and unpredictable, when a large number have passed through the apparatus a pattern is formed on the screen whose intensity distribution is proportional to the intensity of the associated wave. That is, the number of particles arriving at a particular point per unit time is proportional to the square of the amplitude of the wave at that point. It follows that if we apply these ideas to matter waves and consider one particle, the probability that it will be found in a particular place may well be proportional to the square of the modulus of the wave function there. Thus, if $P(x, t)\, dx$ is the probability that the particle is at a point between x and $x + dx$ at a time t, then $P(x, t)$ is proportional to $|\Psi(x, t)|^2$. Thus if we know the wave function associated with a physical system, we can calculate the probability of finding a particle in the vicinity of a particular point. It is a fundamental principle of quantum mechanics that this probability distribution represents all that can be predicted about the particle position: in contrast to classical mechanics which assumes that the position of a particle is always known, or at least knowable, quantum mechanics states that it is almost always uncertain and indeterminate. We shall discuss this indeterminacy in more detail in Chapter 4 where we shall extend the above argument to obtain expressions for the probability distributions governing the measurement of other physical properties, such as the particle momentum, and see how these ideas relate to the uncertainty principle. It is this 'probabilistic' aspect of quantum mechanics which has given rise to many of the conceptual difficulties associated with the subject, and we shall discuss some of these in Chapter 11.

We can now impose an important constraint on the wave function: at any time the particle must certainly be somewhere, so the total probability of finding the particle with an x coordinate between plus and minus infinity must be unity. That is,

$$\int_{-\infty}^{\infty} P(x, t)\, dx = 1 \qquad (2.11)$$

Now, referring back to (2.10), we see that if Ψ is a solution to the Schrödinger equation then $C\Psi$ is also a solution where C is any constant (a differential equation with the properties is said to be *linear*). The scale of the wave function can therefore always be chosen to ensure that the condition (2.11) holds and at the same time

$$P(x, t) = |\Psi(x, t)|^2 \qquad (2.12)$$

This process is known as *normalization*, and a wave function which obeys these conditions is said to be *normalized*. The phase of C, however, is not determined by the normalizing process, and it turns out that a wave function can always be multiplied by a phase factor of the form $\exp(i\alpha)$, where α is an arbitrary, real constant, without affecting the values of any physically significant quantities.

2.2 THE TIME-INDEPENDENT SCHRÖDINGER EQUATION

We consider now the case where the potential, V, is not a function of time and where, according to classical mechanics, energy is conserved. Much of this book will relate to the quantum mechanics of such closed systems and we shall discuss the more general problem of time dependence in detail only in Chapter 8. If V is time independent we can apply the standard 'separation of variables' technique to the Schrödinger equation, putting

$$\Psi(x, t) = u(x)T(t) \tag{2.13}$$

Substituting (2.13) into (2.10) and dividing both sides by Ψ, we get

$$i\hbar \frac{1}{T} \frac{dT}{dt} = -\frac{1}{u} \frac{\hbar^2}{2m} \frac{d^2 u}{dx^2} + V(x) \tag{2.14}$$

Now the left-hand side of this equation is independent of x while the right-hand side is independent of t, but the equation must be valid for all values of x and t. This can be true only if both sides are equal to a constant which we call E. Thus

$$i\hbar \frac{dT}{dt} = ET \tag{2.15}$$

and

$$-\frac{\hbar^2}{2m} \frac{d^2 u}{dx^2} + V(x)u = Eu \tag{2.16}$$

Equation (2.15) can be solved immediately leading to

$$T = \exp(-iEt/\hbar) \tag{2.17}$$

while the solutions to (2.16) depend on the particular form of $V(x)$. This latter equation is known as the one-dimensional *time-independent Schrödinger equation*. In the special case of a free particle, the origin of potential energy can be chosen so that $V(x) = 0$ and a solution to (2.16) is then

$$u = A \exp(ikx)$$

where $k = (2mE/\hbar^2)^{1/2}$ and A is a constant. Thus the wave function has the form

$$\psi = A \exp[i(kx - \omega t)] \tag{2.18}$$

where $\omega = E/\hbar$. This is just the same plane-wave form which we had originally in the case of a free particle (2.2)—provided that the constant E is interpreted as the total energy of the system.†

In the case of any closed system, therefore, we can obtain solutions to the time-dependent Schrödinger equation corresponding to a given value of the energy of the system by solving the appropriate time-independent equation and multiplying the solution by the time-dependent phase factor (2.17). Provided the energy of the system is now known and remains constant (and it is only this case which we shall be considering for the moment) the phase factor, T, has no physical significance. In particular, we notice that the

---

†There are particular difficulties associated with the normalization of a wave function which has a form such as (2.18) and these are discussed in detail in Chapter 9.

probability distribution, $|\Psi|^2$, is now identical to $|u|^2$ and that the normalization condition (2.11) becomes

$$\int_{-\infty}^{\infty} |u|^2 \, dx = 1 \tag{2.19}$$

We shall shortly proceed to obtain solutions to the time-independent Schrödinger equation for a number of forms of the potential, $V(x)$, but before doing so we must establish some boundary conditions that have to be satisfied if the solutions to the Schrödinger equation are to represent physically acceptable wave functions.

2.3 BOUNDARY CONDITIONS

Besides being a solution to the time-independent Schrödinger equation and fulfilling the normalization condition, the wave function must obey the following boundary conditions.

1. The wave function must be a continuous, single-valued function of position and time.

   This boundary condition ensures that the probability of finding a particle in the vicinity of any point is unambiguously defined, rather than having two or more possible values as would be the case if the probability distribution $|\Psi|^2$ were a many-valued function of x (such as $\sin^{-1} x$, for example) or had discontinuities. Although, strictly speaking, this argument only requires $|\Psi|^2$ to be single valued, imposing the condition on the wave function itself ensures the successful calculation of other physical quantities†; an example of this occurs in the discussion of spherically symmetric systems in Chapter 3.

2. The integral of the squared modulus of the wave function over all values of x must be finite.

   In the absence of this boundary condition, the wave function clearly could not be normalized and the probabilistic interpretation would not be possible. We use this condition to reject as physically unrealistic, solutions to the Schrödinger equation that are zero everywhere or which diverge to infinity at any point. A slight modification of this boundary condition and the procedure for normalizing the wave function is necessary in the case of free particles, and this is discussed in Chapter 9.

3. The first derivative of the wave function with respect to x must be continuous everywhere except where there is an infinite discontinuity in the potential.

---

†A detailed discussion of this point has been given by E. Merzbacher, *Am. J. Phys.*, vol. 30, p. 237, 1962.

This boundary condition follows from the fact that a finite discontinuity in $\partial\Psi/\partial x$ implies an infinite discontinuity in $\partial^2\Psi/\partial x^2$ and therefore, from the Schrödinger equation, in $V(x)$.

Having set up these boundary conditions we are now ready to consider the solutions to the Schrödinger equation in some particular cases.

2.4 EXAMPLES

(i) *The Infinite Square Well* As a first example we consider the problem of a particle in the potential $V(x)$ that is illustrated in Fig. 2.1 and is given by

$$V = 0 \qquad -a \leqslant x \leqslant a \qquad (2.20)$$

$$V = \infty \qquad |x| > a \qquad (2.21)$$

This is known as an *infinite square well*.

In the first region, the time-independent Schrödinger equation (2.16) becomes

$$\frac{\hbar^2}{2m}\frac{d^2u}{dx^2} + Eu = 0 \qquad (2.22)$$

The general solution to this equation is well known and can be verified by substitution. It can be written in the form

$$u = A \cos kx + B \sin kx \qquad (2.23)$$

where A and B are constants and $k = (2mE/\hbar^2)^{1/2}$.

In the region outside the well where the potential is infinite, the Schrödinger equation can be satisfied only if the wave function is everywhere zero. We now apply the first boundary condition which requires the wave function to be continuous at $x = \pm a$ and therefore both to be equal to zero and to satisfy (2.23) at these points. Thus

and
$$\left.\begin{array}{l} A \cos ka + B \sin ka = 0 \\ A \cos ka - B \sin ka = 0 \end{array}\right\} \qquad (2.24)$$

Hence, either

$$\left.\begin{array}{llll} B = 0 & \text{and} & \cos ka = 0 & \\ & \text{that is,} & k = n\pi/2a & n = 1, 3, 5, \ldots \\ A = 0 & \text{and} & \sin ka = 0 & \\ & \text{that is,} & k = n\pi/2a & n = 2, 4, 6, \ldots \end{array}\right\} \qquad (2.25)$$

or

These conditions, combined with the definition of k following (2.23), lead

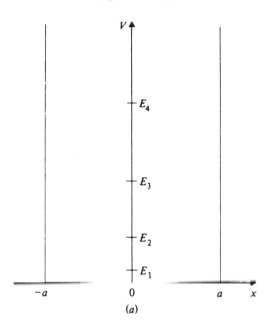

(a)

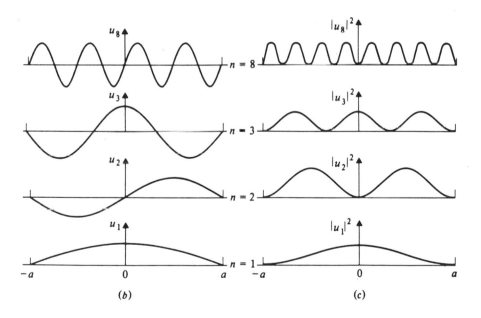

Figure 2.1 (a) shows the potential V as a function of x for an infinite square well, along with energy levels of the four lowest energy states. The wave functions and position probability distributions corresponding to energy states with $n = 1, 2, 3,$ and 8 are shown in (b) and (c) respectively.

to the quantization of the total energy E, according to

$$E \equiv E_n = \hbar^2\pi^2 n^2/8ma^2 \qquad (2.26)$$

Application of the normalization condition (2.19) leads to the following expressions for the time-independent part of the wave function, which we now write as u_n

$$\left.\begin{array}{lll} u_n = a^{-1/2}\cos(n\pi x/2a) & \text{for} & n \text{ odd} \\ u_n = a^{-1/2}\sin(n\pi x/2a) & \text{for} & n \text{ even} \end{array}\right\} \text{ if } -a \leqslant x < a$$

and $\quad u_n = 0 \qquad\qquad\qquad\qquad\qquad\qquad \text{if } |x| > a$

$$(2.27)$$

These expressions are illustrated graphically in Fig. 2.1b for a number of values of n. We see that the wave function is either symmetric ($u_n(x) = u_n(-x)$) or antisymmetric ($u_n(x) = -u_n(-x)$) about the origin, depending on whether n is even or odd. This property is known as the *parity* of the wave function: symmetric wave functions are said to have even parity while antisymmetric wave functions are said to have odd parity. The possession of a particular parity is a general feature of the wave function associated with a particle in a particular energy state of a potential which is itself symmetric (i.e., when $V(x) = V(-x)$).

Remembering that the probability distribution for the particle position is given by $|u(x)|^2$, we see from Fig. 2.1 that, in the lowest energy state, the particle is most likely to be found near the centre of the box, while in the first excited state its most likely positions are near $x = \pm a/2$. For states of comparatively high energy, the probability distribution has the form of a large number of closely spaced oscillations of equal amplitude.

We can use these results to get some idea of how the Schrödinger equation can be used to explain atomic properties. The typical size of an atom is around 10^{-10} m and the mass of an electron is 9.1×10^{-31} kg. Substituting these values into (2.26) leads to the expression

$$E_n \simeq 1.5 \times 10^{-18}n^2 \text{ J}$$

The energy difference between the first and second levels is then 4.5×10^{-18} J (28 eV) so that a photon emitted in a transition between these levels would have a wavelength of about 4.4×10^{-7} m which is of the same order as that observed in atomic transitions. If we perform a similar calculation with m now the mass of a proton (1.7×10^{-27} kg) and a the order of the diameter of a typical nucleus (2×10^{-15} m) the energy difference between the first and second levels is now 5×10^{-12} J (30 MeV) which is in order-of-magnitude agreement with experimental measurements of nuclear binding energies. Of course, neither the atom nor the nucleus is a one-dimensional box, so we can only expect approximate agreement at this stage; quantitative calculations of atomic and nuclear energy levels must wait until we develop a full three-dimensional model in the next chapter.

One of the important requirements of a theory of microscopic systems is that it must produce the same results for macroscopic systems as are successfully predicted by classical mechanics. This is known as the *correspondence principle* and we can test it in the present example by considering a particle of mass 10^{-10} kg (e.g., a small grain of salt) confined to a box of half-width 10^{-6} m. These quantities are small on a macroscopic scale although large in atomic terms. The quantum states of this system then have energies

$$E_n = 1.4 \times 10^{-46} n^2 \text{ J}$$

The minimum energy such a system could possess would be that corresponding to the thermal energy associated with a single degree of freedom. Even at a temperature as low as 1 K this is of the order of 10^{-23} J leading to a value for n of around 3×10^{11}. The separation between adjacent energy levels would then be 8×10^{-35} J and an experiment of the accuracy required to detect any effects due to energy quantization would be completely impossible in practice. At this value of n the separation between adjacent peaks in the probability distribution would be 3×10^{-18} m and an experiment to define the position of the particle to this accuracy or better would be similarly impossible.† Thus to all intents and purposes, quantum and classical mechanics predict the same results—all positions within the well are equally likely and any value of the energy is allowed—and the correspondence principle is verified in this case.

(ii)*The Finite Square Well* We now consider the problem where the sides of the well are not infinite, but consist of finite steps. The potential, illustrated in Fig. 2.2, is then given by

$$\left. \begin{array}{ll} V = 0 & -a \leqslant x \leqslant a \\ V = V_0 & |x| > a \end{array} \right\} \tag{2.28}$$

We shall consider only bound states where the total energy E is less than V_0. The general solution to the Schrödinger equation in the first region is identical to the corresponding result in the infinite case (2.23). In the region $|x| > a$, however, the Schrödinger equation now becomes

$$\frac{\hbar^2}{2m} \frac{d^2u}{dx^2} - (V_0 - E)u = 0 \tag{2.29}$$

whose general solution is

$$u = C \exp(\kappa x) + D \exp(-\kappa x) \tag{2.30}$$

---

† If the energy of the system is not precisely defined then the exact value of n will be unknown. It will be shown later (Chapter 4) that this implies that the wave function is then a linear combination of the wave functions of the states within the allowed energy span. The corresponding probability distribution is then very nearly uniform across the well—in even better agreement with the classical result.

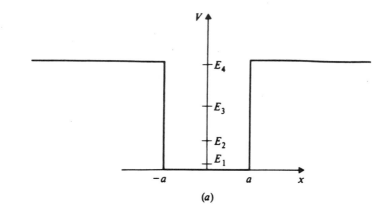

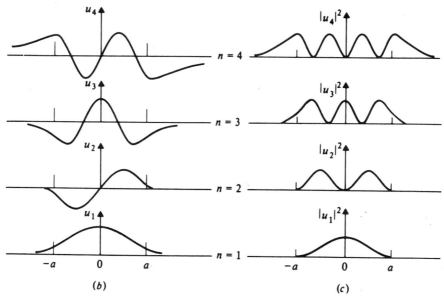

Figure 2.2 (*a*) shows the potential V as a function of x for a finite square well in the case where $V_0 = 25\hbar^2/2ma^2$, along with the energies of the four bound states. The wave functions and position probability distributions for these states are shown in (*b*) and (*c*) respectively.

where C and D are constants and $\kappa = [2m(V_0 - E)/\hbar^2]^{1/2}$. We see at once that C must equal zero otherwise the wave function would tend to infinity as x tends to infinity in breach of the boundary conditions. Thus we have

$$u = D \exp(-\kappa x) \qquad x > a \qquad (2.31)$$

A similar argument leads to

$$u = C \exp(\kappa x) \qquad x < -a \qquad (2.32)$$

As the discontinuities in the potential at $x = \pm a$ are now finite rather than infinite, the boundary conditions require that both u and du/dx be continuous at these points. Thus we have, from (2.23), (2.31), and (2.32),

$$A \cos ka + B \sin ka = D \exp(-\kappa a) \tag{2.33}$$

$$-kA \sin ka + kB \cos ka = -\kappa D \exp(-\kappa a) \tag{2.34}$$

$$A \cos ka - B \sin ka = C \exp(-\kappa a) \tag{2.35}$$

$$kA \sin ka + kB \cos ka = \kappa C \exp(-\kappa a) \tag{2.36}$$

These equations lead directly to

$$2A \cos ka = (C + D) \exp(-\kappa a) \tag{2.37}$$

$$2kA \sin ka = \kappa(C + D) \exp(-\kappa a) \tag{2.38}$$

$$2B \sin ka = (D - C) \exp(-\kappa a) \tag{2.39}$$

$$2kB \cos ka = -\kappa(D - C) \exp(-\kappa a) \tag{2.40}$$

where (2.37) is obtained by adding (2.33) and (2.35), (2.38) is obtained by subtracting (2.34) from (2.36), and (2.39) and (2.40) are derived similarly. If we now divide (2.38) by (2.37) and (2.40) by (2.39) we get

$$\begin{aligned} &\quad k \tan ka = \kappa \qquad \text{unless } C = -D \qquad \text{and} \qquad A = 0 \\ \text{and} \quad &\quad k \cot ka = -\kappa \qquad \text{unless } C = D \qquad \text{and} \qquad B = 0 \end{aligned} \Bigg\} \tag{2.41}$$

The conditions (2.41) and (2.42) must be satisfied simultaneously, so we have two sets of solutions subject to the following conditions:

$$\begin{aligned} \text{either} \quad &k \tan ka = \kappa \qquad C = D \qquad \text{and} \qquad B = 0 \\ \text{or} \quad &k \cot ka = -\kappa \qquad C = -D \qquad \text{and} \qquad A = 0 \end{aligned} \Bigg\} \tag{2.42}$$

These, along with the definitions of k and κ, determine the energy levels and associated wave functions of the system.

Remembering that $k = (2mE)^{1/2}/\hbar$ and $\kappa = [2m(V_0 - E)]^{1/2}/\hbar$, we see that the only unknown quantity in equations (2.41) is E. It follows that these equations determine the allowed values of the energy, just as the energy levels of the infinite well were determined by equations (2.25). However, in the present case the solutions to the equations cannot be expressed as algebraic functions and we have to solve them numerically. One way of doing this is to use the definitions of k and κ to rewrite equations (2.41) as

$$\begin{aligned} &\quad k^2 a^2 \tan^2(ka) = (k_0^2 - k^2)a^2 \\ \text{and} \quad &\quad k^2 a^2 \cot^2(ka) = (k_0^2 - k^2)a^2 \end{aligned} \Bigg\} \tag{2.43}$$

where $k_0^2 = 2mV_0/\hbar^2$, which in turn can be rewritten using standard trigonometric identities as

$$
\left.
\begin{aligned}
ka &= n_1\pi + \cos^{-1}(ka/k_0a) \\
ka &= n_2\pi - \sin^{-1}(ka/k_0a)
\end{aligned}
\right\}
\qquad (2.44)
$$

and

where n_1 and n_2 are integers and the terms $n_1\pi$ and $n_2\pi$ are included because of the multivalued property of the inverse cosine and sine functions. In general, solutions will exist for several values of n_1 and n_2 corresponding to the different energy levels. However, it is clear that solutions do not exist if $n_1\pi$ or $n_2\pi$ is appreciably greater than k_0a because the arguments of the inverse cosine or sine would then have to be greater than one. This corresponds to the fact that there is a limited number of bound states with energies less than V_0.

Values for ka and hence E can be obtained by straightforward iteration. First, we evaluate k_0a from the values of V_0 and a for the particular problem. If we now guess a value for ka, we can substitute this into the right-hand side of one of (2.44) and obtain a new value of ka. This process usually converges to the correct value of ka. However, if the required value of ka is close to k_0a, iteration using (2.44) can fail to converge. Such cases can be successfully resolved by applying a similar iterative process to the equivalent equations

$$
\left.
\begin{aligned}
ka &= k_0a \cos(ka - n_1\pi) \\
ka &= k_0a \sin(n_2\pi - ka)
\end{aligned}
\right\}
\qquad (2.44a)
$$

and

The reader should try this for the case where $V_0 = 25\hbar^2/2ma^2$ so that k_0a equals 5.0. The ground state energy can be obtained from the first of (2.44) with $n_1 = 0$; starting with an initial value of ka anywhere between 1.0 and 2.0, ka should converge to 1.306 after a few iterations. If the exercise is repeated with $n_1 = 1$, another solution with $ka = 3.838$ should be obtained. However, if we try $n_1 = 2$, we are unable to obtain a solution, because the energy would now be greater than V_0. The remaining levels can be found by a very similar procedure using the second of equations (2.44) and (2.44a). Table 2.1 sets out the details of all the possible solutions in this case, showing the energy levels both as fractions of V_0 and as fractions of the energies of the corresponding infinite-well states (2.26). The associated wave functions are shown in Fig. 2.2. Comparing these with the wave functions for the infinite square well (Fig. 2.1), we see that they are generally similar and, in particular, that they have a definite parity, being either symmetric or antisymmetric about the point $x = 0$. However, one important difference between Figs 2.2 and 2.1 is that in the former case the wave functions decay exponentially in the region $|x| > a$ instead of going to zero at $x = \pm a$. That is, the wave function penetrates a region where the total energy is less than V_0, implying that there is a probability of finding the particle in a place where

Table 2.1 Values of the quantities ka, κa, and E that are consistent with the boundary conditions for a potential well whose sides are of height V_0 when $V_0 = 25\hbar^2/2ma^2$. The energies of the corresponding states in the case where V_0 is infinite are represented by E_∞.

ka	κa	E/V_0	E/E_∞
1.306	4.826	0.069	0.691
2.596	4.273	0.270	0.682
3.838	3.205	0.590	0.663
4.907	0.960	0.964	0.610

it could not be classically as it would then have to have negative kinetic energy. This is another example of a quantum-mechanical result that is quite different from the classical expectation and we shall discuss it in more detail in the next section.

The penetration of the wave function into the classically forbidden region also results in the energy levels being lower than in the infinite square-well case (Table 2.1) because the boundary conditions are now satisfied for smaller values of k. This effect is more noticeable for the higher energy levels and, conversely, we can conclude that in the case of a very deep well, the energy levels and wave functions of the low-lying states would be indistinguishable from those where V_0 was infinite. This point also follows directly from the boundary conditions: when $(V_0 - E)$ and therefore κ are very large, the conditions (2.43) and (2.44) become identical to (2.25).

2.5 QUANTUM MECHANICAL TUNNELLING

We now turn to a more detailed discussion of effects associated with the penetration of the wave function into the classically forbidden region. We first consider a potential well bounded by barriers of finite height and width as in Fig. 2.3a. Although it is possible to obtain solutions to the Schrödinger equation for this case that are consistent with the boundary conditions, this gives rise to considerable algebraic complication and it will be sufficient to discuss the problem qualitatively, making use of the results of the earlier examples. Firstly the low-lying energy states (those with $E \ll V_1$) will have wave functions similar to those of the finite square well with a small amount of penetration into the regions where $|x|$ is greater than a. States with energy greater than V_1 and less than V_2 have particularly interesting properties. In

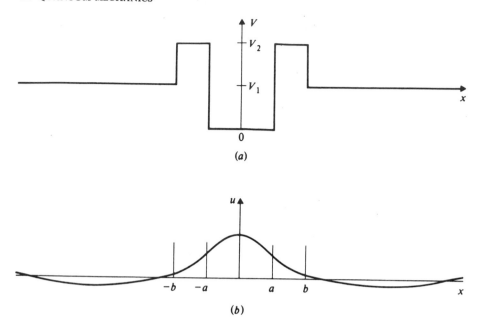

Figure 2.3 The wave function, corresponding to one of the bound states of the potential shown in (a), is illustrated in (b).

the central region the wave function is sinusoidal with $k = (2mE/\hbar^2)^{1/2}$ as before, and in the classically forbidden region the wave function decays exponentially in a manner similar to that seen in the finite square well. In the region $|x| > b$, however, the total energy is again greater than the potential energy and the wave function is again oscillatory, although now with a smaller value of k (equal to $[2m(E - V_1)/\hbar^2]^{1/2}$) than when $|x| < a$. When the wave function and its derivatives are matched at the different boundaries, a solution of the form shown in Fig. 2.3b is obtained, from which we see that there is a probability of finding the particle both inside and outside the potential well and also at all points within the barrier. Quantum mechanics therefore implies that a particle is able to pass through a potential energy barrier which, according to classical mechanics, should be impenetrable. This phenomenon is known as *quantum mechanical tunnelling* or *the tunnel effect*. The probability of such tunnelling is proportional to the ratio of the probabilities of finding the particle on either side of the barrier, i.e. to the ratio of the squares of the wave function at these points. Unless the barrier is very narrow, this quantity can be calculated assuming that the wave function decays under the barrier in the same way as it does within the potential step discussed above. It follows from (2.31) that this probability ratio is $\exp[-2\kappa(b - a)]$ where $\kappa = [2m(V_0 - E)]^{1/2}/\hbar$ as before.

A number of physical examples of tunnelling have been observed and two of these—alpha particle decay and cold electron emission—will now be described.

Alpha Decay

It is well known that some nuclei decay radioactively emitting alpha particles. The alpha particle consists of two protons and two neutrons bound together so tightly that it can be considered as retaining this identity even when within the nucleus. The interaction between the alpha particle and the rest of the nucleus is made up of two components; the first results from the so-called strong nuclear force which is attractive, but of very short range, whereas the second is the Coulomb interaction which is repulsive (because both the alpha particle and the residual nucleus are positively charged) and acts at comparatively large distances. The total interaction potential energy is sketched in Fig. 2.4a as a function of the separation between the alpha

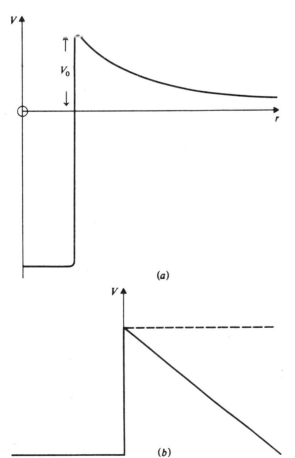

(a)

(b)

Figure 2.4 (a) shows the potential energy of interaction between an α particle and a nucleus as a function of its distance from the centre of the nucleus while (b) shows the potential energy of an electron near the surface of a metal with and without (broken line) an applied electric field. In each case the particles can pass through the potential barrier by quantum-mechanical tunnelling.

particle and the nucleus, and we see that it is qualitatively similar to that shown in Fig. 2.3 and discussed above. It follows that if the alpha particle occupies a quantum state whose energy is less than zero, it will remain there indefinitely and the nucleus will be stable. If, on the other hand, the form of the potential is such that the lowest energy state of the alpha particle is greater than zero, but less than V_0, it will be able to escape from the nucleus by quantum-mechanical tunnelling. The probability of such emission will depend on the actual shape of the barrier, particularly its height and width, which accounts for the large variation in the observed decay constants of different nuclei.

Cold Electron Emission

This phenomenon is observed when a strong electric field is directed towards the surface of a metal, resulting in the emission of electrons. This occurs even if the electrons are not thermally excited (which would be thermionic emission) and so do not have enough energy to escape classically. We first consider the situation in the absence of a field (Fig. 2.4b); the electrons are then confined within the metal by an energy barrier formed by the work function (see the discussion of the photoelectric effect in Chapter 1). When the electric field is applied, the potential is changed, so that at a short distance from the surface of the metal the potential energy is less than the energy of the electrons inside the metal. Now, although the electrons cannot classically penetrate the barrier at the metal surface, they can pass through by quantum-mechanical tunnelling and the observation of cold electron emission is therefore a confirmation of this effect. In recent years, cold electron emission has been exploited in the *scanning tunnelling microscope*. In this instrument, an electric potential is maintained between a very sharp tungsten point and a metal surface above which it is held very closely. A tunnelling current between the surface and the point is measured and the point is scanned slowly across the metal surface. Variations in the tunnelling current then represent changes in the separation between the point and the source of electron emission. In this way it is possible to observe the changes as the point moves over individual atoms and so to map the actual atomic structure on the metal surface.

It should be noted that, although both the above experiments imply that the alpha particle or electron has passed through a classically forbidden region, in neither case has the particle been directly observed while undergoing this process. In fact such an observation appears always to be impossible; all known particle detectors are sensitive only to particles with positive kinetic energy so that, if we insert such a detector within the classically forbidden region, its presence implies that a 'hole' has been made in the potential so that the particle is no longer in such a region when it is detected.

2.6 THE HARMONIC OSCILLATOR

We finish this chapter with a discussion of the energy levels and wave functions of a particle moving in a harmonic oscillator potential. This is an important example because many physical phenomena, including the internal vibrations of molecules and the motion of atoms in solids, can be described using it. It also provides an application of the method of series solution of differential equations, which is a technique that will be widely used in the next chapter when we consider three-dimensional systems.

The harmonic oscillator potential has the form $V(x) = \frac{1}{2}Kx^2$ where K is a constant, and classically a particle of mass m oscillates in this potential with an angular frequency $\omega_c = (K/m)^{1/2}$. The time-independent Schrödinger equation for this system can therefore be written in the form

$$-\frac{\hbar^2}{2m}\frac{d^2u}{dx^2} + \frac{1}{2}m\omega_c^2 x^2 u = Eu \qquad (2.45)$$

The subsequent mathematics is a good deal easier to follow if we first change variables from x to y, where $y = (m\omega_c/\hbar)^{1/2}x$ and define a constant $\alpha = (2E/\hbar\omega_c)$. Equation (2.45) now becomes

$$\frac{d^2u}{dy^2} + (\alpha - y^2)u = 0 \qquad (2.46)$$

We first discuss the asymptotic form of solution in the region where y is very large so that the equation is approximately

$$\frac{d^2u}{dy^2} - y^2u = 0 \qquad (2.47)$$

We shall try as a solution to this equation

$$u = y^n \exp(-y^2/2) \qquad (2.48)$$

where n is a positive integer. Differentiating twice with respect to y we get

$$\frac{d^2u}{dy^2} = [n(n-1)y^{n-2} - (2n+1)y^n + y^{n+2}]\exp(-y^2/2)$$

$$\simeq y^{n+2}\exp(-y^2/2) \qquad \text{when } y \gg 1$$

$$= y^2u$$

Thus we see that (2.48) is the asymptotic form of solution we are looking for, which suggests that a general solution to (2.46), valid for all values of y, might be

$$u(y) = H(y)\exp(-y^2/2) \qquad (2.49)$$

where $H(y)$ is a function to be determined. Substituting from (2.49) into

(2.46) we get

$$H'' - 2yH' + (\alpha - 1)H = 0 \qquad (2.50)$$

where a prime indicates differentiation with respect to y.

We now write H in the form of a power series:

$$H = \sum_{p=0}^{\infty} a_p y^p \qquad (2.51)$$

(Note that negative powers of y are not permitted as they produce physically unacceptable infinities at $y = 0$.) Hence

$$H' = \sum_{p=0}^{\infty} a_p p y^{p-1} \qquad (2.52)$$

and

$$H'' = \sum_{p=0}^{\infty} a_p p(p-1) y^{p-2}$$
$$= \sum_{p=2}^{\infty} a_p p(p-1) y^{p-2} \qquad (2.53)$$

because the first two terms on the right-hand side of (2.53) vanish; thus,

$$H'' = \sum_{p=0}^{\infty} a_{p+2}(p+2)(p+1) y^p \qquad (2.54)$$

where we have now redefined p (which is just an index of summation) as its previous value plus two. Substituting from (2.51), (2.52), and (2.54) into (2.50) we get

$$\sum_{p=0}^{\infty} [(p+1)(p+2)a_{p+2} - (2p+1-\alpha)a_p] y^p = 0 \qquad (2.55)$$

This can be true for all values of x only if the coefficient of each power of y vanishes, so we obtain the following recurrence relation:

$$a_{p+2}/a_p = (2p+1-\alpha)/[(p+1)(p+2)]$$
$$\rightarrow 2/p \quad \text{as } p \rightarrow \infty \qquad (2.56)$$

This last expression is identical to the recurrence relation between successive terms of the power series for the function $\exp(y^2)$ $(= \sum_n (y^{2n}/n!))$ so in general $H(y)$ will tend to infinity with y like $\exp(y^2)$ and hence $u(y)$ will diverge like $\exp(\frac{1}{2}y^2)$ leading to a physically unrealistic solution (cf. **2.3**). This can be avoided only if the power series for H terminates after a finite number of terms. To obtain the conditions for such a termination, we first note that the series can be expressed as a sum of two series, one containing only even and the other only odd powers of y. By repeated application of (2.56) we can express the coefficients of y^p as functions of α multiplied by the constants a_0 or a_1 depending on whether p is even or odd, respectively. Either series, but not both, can be made to terminate by choosing α so that the numerator of (2.56) vanishes for some finite value of p (say $p = n$)—i.e., by putting α equal to $2n + 1$. The other series cannot terminate simultaneously, but can be made to vanish completely if its leading coefficient (a_0 or a_1) is taken to be zero. Thus we have the following conditions for a physically acceptable solution to the Schrödinger equation in the case of a particle moving in a harmonic oscillator potential:

(i) $\alpha = 2n + 1$ $n = 0, 1, 2, \ldots$

(ii) $a_1 = 0$ if n is even   and $a_0 = 0$ if n is odd   (2.57)

If the first condition is combined with the definition of α, we find that the total energy of the system is quantized according to

$$E = E_n = (n + \tfrac{1}{2})\hbar\omega_c \tag{2.58}$$

Thus quantum mechanics predicts that the energy levels of a harmonic oscillator are equally spaced with an interval of $\hbar$ times the classical frequency and have a minimum value of $\tfrac{1}{2}\hbar\omega_c$ (known as the *zero-point energy*). Experimental confirmation of these results is obtained, for example, from observations of the properties of molecules. Thus a diatomic molecule can be considered as two point masses connected by a spring and this system can therefore undergo quantized oscillations as described above; photons absorbed and emitted by such a molecule then have frequencies that are multiples of the classical frequency of the oscillator. Thermal properties, like the heat capacity, of a gas composed of such molecules are also significantly affected by energy quantization.

The polynomials $H_n(y)$ are known as Hermite polynomials. Expressions for them can be obtained by repeated application of (2.56) and the results of this procedure for the four lowest energy states are:

$$\left. \begin{aligned} H_0 &= 1 \\ H_1 &= 2y \\ H_2 &= 4y^2 - 2 \\ H_3 &= 8y^3 - 12y \end{aligned} \right\} \tag{2.59}$$

where the values of a_0 and a_1 have been chosen according to established convention. The wave functions $u_n(x)$ can now be obtained by multiplying $H_n(y)$ by the factor $\exp(-\tfrac{1}{2}y^2)$, making the substitution $y = (m\omega_c/\hbar)^{1/2}x$ and applying the normalization condition (2.19). The results of this procedure are shown in Fig. 2.5 where the wave functions are seen to have even or odd parity, depending on whether n is even or odd. We note that, like the wave functions associated with the energy states of a finite square well (Fig. 2.2),

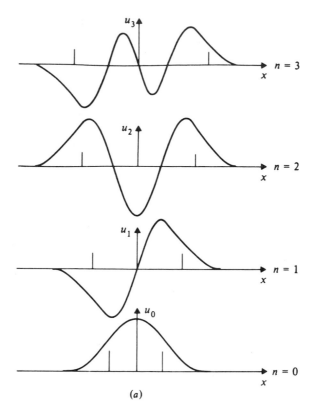

(a)

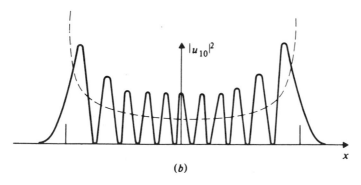

(b)

Figure 2.5 The wave functions corresponding to the four lowest energy states of the harmonic oscillator are shown in (a) while (b) shows the position probability distribution in the case where $n = 10$, compared with that calculated classically (broken line). The limits of the classical oscillation are indicated by vertical lines in each case.

they are oscillatory within the classically permitted region and penetrate the classically forbidden region to some extent. Algebraic expressions for these wave functions are given below:

$$\left.\begin{aligned} u_0 &= (m\omega_c/\pi\hbar)^{1/4} \exp(-m\omega_c x^2/2\hbar) \\ u_1 &= (4/\pi)^{1/4}(m\omega_c/\hbar)^{3/4} x \exp(-m\omega_c x^2/2\hbar) \\ u_2 &= (m\omega_c/4\pi\hbar)^{1/4}[2(m\omega_c/\hbar)x^2 - 1]\exp(-m\omega_c x^2/2\hbar) \\ u_3 &= (1/9\pi)^{1/4}(m\omega_c/\hbar)^{3/4}[(2m\omega_c/\hbar)x^2 - 3]x \exp(-m\omega_c x^2/2\hbar) \end{aligned}\right\}(2.60)$$

The harmonic oscillator provides yet another example of the correspondence principle whereby the results of quantum mechanics tend to those of classical mechanics in the classical limit. To see this, we consider a particle of mass 10^{-10} kg oscillating with a classical angular frequency $\omega_c = 10^6$ rad s^{-1}. The separation of the energy levels is then 10^{-28} J which corresponds to a temperature of about 10^{-5} K. At any normal temperature, therefore, the oscillator will be highly excited and the effects of energy quantization will be undetectable. We can also show that the probability distribution of the particle's position is similar to that expected classically if the oscillator is highly excited. Classically the position and velocity of an oscillating particle vary as

$$x = a \cos \omega_c t \qquad v = -a\omega \sin \omega_c t \qquad (2.61)$$

where a is the amplitude of oscillation. Now an oscillating particle is more likely to be found near the limits of its oscillation where it is moving slowly than near the centre where it moves more quickly. Putting this point more quantitatively, the probability $P(x)\, dx$ that the particle be found in the region between x and dx is clearly inversely proportional to the speed $v(x)$ at the point x. Hence using (2.61)

$$\left.\begin{aligned} P(x) &= \pi/(a^2 - x^2)^{1/2} & |x| < a \\ &= 0 & |x| > a \end{aligned}\right\}(2.62)$$

where the factor π ensures normalization. It is clear from Fig. 2.5a that the quantum-mechanical probability distribution is quite different from this for low energy states, but the agreement rapidly improves as n increases. A detailed comparison is shown for the case $n = 10$ in Fig. 2.5b. This shows that the quantum-mechanical probability distribution is a rapidly oscillating function of x, but that if we average over these oscillations, the result is quite close to (2.62). For larger values of n this agreement is even better and the separation of the adjacent maxima and minima in the quantum-mechanical probability distribution also becomes undetectably small so that the predictions of quantum and classical mechanics are indistinguishable.

PROBLEMS

2.1 An electron is confined to a one-dimensional potential well of width 3×10^{-10} m which has infinitely high sides. Calculate: (i) the three lowest allowed values of the electron energy; (ii) the wavelength of the electromagnetic wave that would cause the electron to be excited from the lowest to the highest of these three levels; (iii) all possible wavelengths of the radiation emitted following the excitation in (ii).

2.2 If u_m and u_n are the wave functions corresponding to two energy states of a particle confined to a one-dimensional box with infinite sides, show that

$$\int_{-\infty}^{\infty} u_n u_m \, dx = 0 \qquad \text{if } n \neq m$$

This is an example of 'orthogonality' which will be discussed in Chapter 4.

2.3 Consider a particle of mass m subject to a one-dimensional potential $V(x)$ that is given by

$$V = \infty, \qquad x < 0; \qquad V = 0, \qquad 0 \leqslant x \leqslant a; \qquad V = V_0, \qquad x > a$$

Show that bound $(E < V_0)$ states of this system exist only if $k \cot ka = -\kappa$ where $k^2 = 2mE/\hbar^2$ and $\kappa^2 = 2m(V_0 - E)/\hbar^2$.

2.4 Show that if $V_0 = 9\hbar^2/2ma^2$, only one bound state of the system described in Prob. 2.3 exists. Calculate its energy as a fraction of V_0 and sketch its wave function, using an iteration similar to that discussed in Section 2.4.

2.5 Show that if $V(x) = V(-x)$, solutions to the time-dependent Schrödinger equation must have definite parity—that is, $u(x) = \pm u(-x)$.
   Hint: Make the substitution $y = -x$ and show first that $u(x) = Au(-x)$ where A is a constant.

2.6 Consider a particle of mass m subject to the one-dimensional potential $V(x)$ that is given by

$$V = 0 \qquad \text{if } -a \leqslant x \leqslant a \qquad \text{or} \qquad \text{if } |x| > b$$
$$V = V_0 \qquad \text{if } a < |x| \leqslant b$$

where $b > a$. Write down the form of an even-parity solution to the Schrödinger equation in each region in the case where $E < V_0$. Note that the particle is not bound in this potential as there is always a probability of quantum-mechanical tunnelling, so solutions exist for all values of E. Show, however, that if $\kappa(b - a) \gg 1$ (where κ is defined as in Prob. 2.3) the probability of finding the particle inside the region $|x| < a$ is very small unless its energy is close to that of one of the bound states of a well of side $2a$ bounded by potential steps of height V_0. In the case where this condition is fulfilled exactly, obtain an expression for the ratio of the amplitudes of the wave function in the regions $|x| \leqslant a$ and $|x| > b$.

2.7 The hydrogen atom in a water molecule can vibrate in a direction along the O–H bond, and this motion can be excited by electromagnetic radiation of a wavelength about 4×10^{-6} m, but not by radiation of a longer wavelength. Calculate the effective spring constant for this vibration and the zero-point energy of the oscillator. Given that every molecular degree of freedom has a thermal energy of about $k_B T$, where k_B (Boltzmann's constant) $\simeq 1.4 \times 10^{-23}$ J K^{-1} and T is the temperature, what is the most probable vibrational state in the case of a water molecule in steam at 450 K?

2.8 Calculate the normalization constants for the two lowest energy states of a harmonic oscillator and verify that they are orthogonal in the sense defined in Prob. 2.2.

THREE

THE THREE-DIMENSIONAL SCHRÖDINGER EQUATIONS

In the previous chapter we saw that for a particle in a one-dimensional potential well, physically acceptable solutions to the one-dimensional Schrödinger equation are possible only for particular values of the total energy; moreover, in the case of an electron in a well of atomic dimensions, the spacings between these energy levels are in qualitative agreement with the separations experimentally observed in atoms. We also showed that if the square of the wave function is interpreted as a probability distribution for the position of the particle, phenomena such as quantum-mechanical tunnelling are predicted which have been confirmed by experiment. The real world, however, is three-dimensional, and, although one-dimensional examples often provide useful insights and analogies, we have to extend our theory into three dimensions before we can make quantitative predictions of most experimental results. In the present chapter, therefore, we shall set up the three-dimensional Schrödinger equation and obtain its solutions in a number of cases, culminating in a discussion of the hydrogen atom where we shall find that theory and experiment agree to a remarkable degree of accuracy.

3.1 THE WAVE EQUATIONS

In classical mechanics the total energy of a free particle of mass m and momentum p is given by

$$E = p^2/2m = (p_x^2 + p_y^2 + p_z^2)/2m \qquad (3.1)$$

where p_x, p_y, and p_z are the momentum components along the Cartesian axes x, y, and z. The de Broglie relations (1.10) in three dimensions become:

$$\left.\begin{array}{l} E = \hbar\omega \\[4pt] \mathbf{p} = \hbar\mathbf{k} \qquad \text{that is} \qquad p_x = \hbar k_x ; p_y = \hbar k_y ; p_z = \hbar k_z \end{array}\right\} \tag{3.2}$$

A wave equation whose solutions are consistent with these relations can be set up in exactly the same way as was described in the one-dimensional case (Eqs (2.1) to (2.8)). We obtain

$$i\hbar \frac{\partial\Psi}{\partial t} = -\frac{\hbar^2}{2m}\left(\frac{\partial^2\Psi}{\partial x^2} + \frac{\partial^2\Psi}{\partial y^2} + \frac{\partial^2\Psi}{\partial z^2}\right) \tag{3.3}$$

where the wave function $\Psi(\mathbf{r}, t)$ is now a function of all three positional coordinates and the time. When the particle is not free, but is subject to a potential $V(\mathbf{r}, t)$, Eq. (3.3) is generalized in the same way as in the one-dimensional case (Eqs (2.9) to (2.10)) to produce the time-dependent Schrödinger equation

$$i\hbar \frac{\partial\Psi}{\partial t} = -\frac{\hbar^2}{2m}\nabla^2\Psi + V\Psi \tag{3.4}$$

where we have used the vector operator ∇^2 (del-squared) which is defined so that

$$\nabla^2\Psi = \frac{\partial^2\Psi}{\partial x^2} + \frac{\partial^2\Psi}{\partial y^2} + \frac{\partial^2\Psi}{\partial z^2}$$

The generalization of the probabilistic interpretation of the wave function is also straightforward: if $P(\mathbf{r}, t)\,d\tau$ is the probability that the particle be found in the volume element $d\tau$ ($\equiv dx\,dy\,dz$) in the vicinity of the point $\mathbf{r}$ at time t, then

$$P(\mathbf{r}, t) = |\Psi(\mathbf{r}, t)|^2 \tag{3.5}$$

It follows directly that the normalization condition (2.12) becomes

$$\int |\Psi(\mathbf{r}, t)|^2 \, d\tau = 1 \tag{3.6}$$

where the volume integral is performed over all space (as will always be assumed to be the case for volume integrals unless we specifically state otherwise).

When the potential V is independent of time, we can write $\Psi(\mathbf{r}, t) = u(\mathbf{r})T(t)$ and separate the variables to obtain the time-independent Schrödinger equation (cf. Eqs (2.13) to (2.17))

$$-\frac{\hbar^2}{2m}\nabla^2 u + Vu = Eu \tag{3.7}$$

along with

$$T(t) = \exp(-iEt/\hbar)$$

while the normalization condition (3.6) now becomes

$$\int |u(\mathbf{r})|^2 \, d\tau = 1 \qquad (3.8)$$

The boundary conditions on the wave function also follow as a natural extension of the arguments set out in Chapter 2: it must be a continuous, single-valued function of position and time; its squared modulus must be integrable over all space—which usually means that it must be everywhere finite; and its spatial derivatives ($\partial\Psi/\partial x$, $\partial\Psi/\partial y$, and $\partial\Psi/\partial z$) must all be continuous everywhere except where there is an infinite discontinuity in V.

We shall now proceed to obtain solutions to the three-dimensional time-independent Schrödinger equation (3.7) in a number of particular cases. Unlike the one-dimensional case, the equation is now a partial differential equation which gives rise to considerable mathematical complications in general and, even though we shall consider only cases where the separation of variables technique can be used, we shall still have to solve three ordinary differential equations in order to obtain a complete solution. We shall shortly consider spherically symmetric systems when we shall carry out this process in a spherical polar coordinate system, but we first discuss some simpler examples where the Schrödinger equation can be separated in Cartesian coordinates.

3.2 SEPARATION IN CARTESIAN COORDINATES

Consider the case where the potential $V(\mathbf{r})$ can be written as a sum of three quantities each of which is a function of only one of the three Cartesian coordinates. That is,

$$V(\mathbf{r}) = V_1(x) + V_2(y) + V_3(z) \qquad (3.9)$$

We now express the wave function as a product of three one-dimensional functions

$$u(\mathbf{r}) = X(x)Y(y)Z(z) \qquad (3.10)$$

On substituting (3.9) and (3.10) into the time-independent Schrödinger equation (3.7), dividing through by u and rearranging slightly, we get

$$\left[-\frac{\hbar^2}{2m} \frac{1}{X} \frac{d^2 X}{dx^2} + V_1(x) \right] + \left[-\frac{\hbar^2}{2m} \frac{1}{Y} \frac{d^2 Y}{dy^2} + V_2(y) \right]$$

$$+ \left[-\frac{\hbar^2}{2m} \frac{1}{Z} \frac{d^2 Z}{dz^2} + V_3(z) \right] = E \qquad (3.11)$$

Each expression in square brackets is a function of only one of the variables (x, y, z) so Eq. (3.11) can be satisfied at all points in space only if each separate function is equal to a constant and if the sum of the constants equals E. Thus

$$\left.\begin{aligned}
-\frac{\hbar^2}{2m}\frac{d^2X}{dx^2} + V_1 X = E_1 X \\
-\frac{\hbar^2}{2m}\frac{d^2Y}{dy^2} + V_2 Y = E_2 Y \\
-\frac{\hbar^2}{2m}\frac{d^2Z}{dz^2} + V_3 Z = E_3 Z
\end{aligned}\right\} \qquad (3.12)$$

where $E_1 + E_2 + E_3 = E$. Each of these equations has the form of the one-dimensional Schrödinger equation (2.16) so, in suitable cases, we can carry over results directly from the previous chapter.

Example 3.1 The three-dimensional 'box' This example relates to a potential which is zero inside a rectangular region of sides $2a \times 2b \times 2c$ and infinite outside. That is, we are considering a potential of the form

$$\left.\begin{aligned}
V(\mathbf{r}) = 0 \qquad &\text{if } -a \leqslant x \leqslant a, -b \leqslant y \leqslant b \text{ and } -c \leqslant z \leqslant c \\
V(\mathbf{r}) = \infty \qquad &\text{if } |x| > a, |y| > b, \text{ or } |z| > c
\end{aligned}\right\} \qquad (3.13)$$

Each of the three separated equations (3.12) is now equivalent to the Schrödinger equation for the one-dimensional infinite square well and the boundary conditions ($X = 0$ if $x = \pm a$, etc.) are also the same. It therefore follows directly from (2.26) that the energy levels are given by

$$E_{n_1 n_2 n_3} = \frac{\hbar^2 \pi^2}{8m}\left(\frac{n_1^2}{a^2} + \frac{n_2^2}{b^2} + \frac{n_3^2}{c^2}\right) \qquad (3.14)$$

where n_1, n_2, and n_3 are integers and the complete wave function is given (cf. (2.27)) by

$$u_{n_1 n_2 n_3} = (abc)^{-1/2}\begin{Bmatrix}\cos\\\sin\end{Bmatrix}\left(\frac{n_1 \pi x}{2a}\right)\begin{Bmatrix}\cos\\\sin\end{Bmatrix}\left(\frac{n_2 \pi y}{2b}\right)\begin{Bmatrix}\cos\\\sin\end{Bmatrix}\left(\frac{n_3 \pi z}{2c}\right) \qquad (3.15)$$

where the cosine applies if the integer in the following argument is odd, and the sine if it is even.

We see that three quantum numbers (n_1, n_2, and n_3) are needed to specify the energy and wave function in this example, compared with only one in the one-dimensional case. This is a general feature of three-dimensional bound systems, as, following the separation of variables, there are three ordinary differential equations to be solved, and each one gives rise to a quantum condition.

It is interesting to consider the special case where two sides of the box are equal, as this illustrates some important features that arise when a three-dimensional potential has symmetry. Putting $a = b$, (3.14) becomes

$$E_{n_1 n_2 n_3} = \frac{\hbar^2 \pi^2}{8m} \left(\frac{n_1^2 + n_2^2}{a^2} + \frac{n_3^2}{c^2} \right) \tag{3.16}$$

There are now, in general, several different combinations of n_1, n_2, and n_3 that have the same energy. For example the states ($n_1 = 2$, $n_2 = 1$, $n_3 = 1$) and ($n_1 = 1$, $n_2 = 2$, $n_3 = 1$) have the same energy while their wave functions are

$$\left. \begin{array}{l} u_{211} = (a^2 c)^{-1/2} \sin(\pi x/a) \cos(\pi y/2a) \cos(\pi z/2c) \\ u_{121} = (a^2 c)^{-1/2} \cos(\pi x/2a) \sin(\pi y/a) \cos(\pi z/2c) \end{array} \right\} \tag{3.17}$$

When two or more quantum states have the same value of the energy we say that they are *degenerate*. Clearly the degeneracy in the present case is closely associated with the symmetry of the potential and we can illustrate this further by considering the geometrical relationship between the two wave functions given in (3.17). Figure 3.1 shows contour diagrams of the section at $z = 0$ through each of these functions and we can see that they are equivalent to each other, apart from their orientation in space: thus u_{121} can be transformed into u_{211} by rotating it through 90° about the z axis. As the potential has square symmetry in the xy plane, we should not expect such a rotation to result in any physical change of the system, and it follows that the degeneracy of the two energy states is a necessary consequence of this symmetry. The position probability distribution, however, does not appear to have the expected symmetry: if the system is in the state with wave function u_{211}, for example, the probability of finding it near $(a/2, 0, 0)$ is quite large, while that of finding it near $(0, a/2, 0)$ is zero—even though, as far as the potential is concerned, these points are symmetrically equivalent. To resolve this apparent paradox we must carefully consider what information a measurement of the energy gives us about such a degenerate system: if no further measurements have been made we can conclude that the wave function is one of the two forms given in (3.17), but we cannot tell which. The appropriate expression for the position probability distribution is therefore the *average* of $|u_{211}|^2$ and $|u_{121}|^2$, and this quantity clearly does have the same symmetry as the potential, as is also shown in Fig. 3.1. In the case of degeneracy, therefore, even the squared moduli of the wave functions associated with the individual states do not have a direct physical significance and, as usual, we see that the predictions of quantum mechanics only make sense provided we concentrate on those results that can be measured and avoid drawing conclusions about apparent consequences which cannot in fact be directly tested.

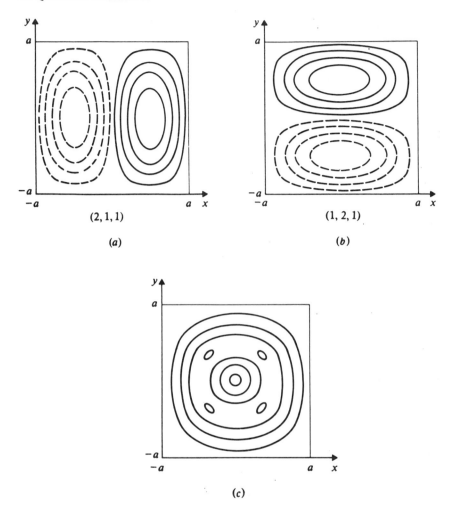

Figure 3.1 Sections at $z = 0$ through the three-dimensional wave functions of a particle in a rectangular box with $a = b$ corresponding to the states (a) $n_1 = 2$, $n_2 = 1$, $n_3 = 1$, and (b) $n_1 = 1$, $n_2 = 2$, $n_3 = 1$. Continuous lines represent positive contours and broken lines represent negative contours. A section at $z = 0$ through the position probability distribution calculated as the average of the squares of (a) and (b) is shown in (c).

We shall return to the topic of degeneracy when we discuss spherically symmetric systems later in this chapter and we shall also consider the topic more formally in Chapter 4.

Example 3.2 The three-dimensional harmonic oscillator As another example of a three-dimensional system where the Schrödinger equation can be solved using Cartesian coordinates, we consider a particle moving in the potential

$$V(\mathbf{r}) = \tfrac{1}{2}K_1 x^2 + \tfrac{1}{2}K_2 y^2 + \tfrac{1}{2}K_3 z^2 \tag{3.18}$$

The separated Schrödinger equations (3.12) are now

$$-\frac{\hbar^2}{2m}\frac{d^2 X}{dx^2} + \tfrac{1}{2}m\omega_1^2 x^2 X = E_1 X \tag{3.19}$$

where $\omega_1 = (K_1/m)^{1/2}$, with similar equations for Y and Z. Each of these has the form of the one-dimensional harmonic oscillator equation (2.45) so we can use the results of this case (2.58) directly to get an expression for the energy levels of the three-dimensional oscillator

$$E_{n_1 n_2 n_3} = (n_1 + \tfrac{1}{2})\hbar\omega_1 + (n_2 + \tfrac{1}{2})\hbar\omega_2 + (n_3 + \tfrac{1}{2})\hbar\omega_3 \tag{3.20}$$

where n_1, n_2, and n_3 are positive integers. The wave functions also follow directly from the one-dimensional results

$$u_{n_1 n_2 n_3} = H_{n_1}(x')H_{n_2}(y')H_{n_3}(z')\exp - \tfrac{1}{2}(x'^2 + y'^2 + z'^2) \tag{3.21}$$

where $x' = (m\omega_1/\hbar)^{1/2}x$, etc., and the H_{n_i}'s are Hermite polynomials.

This example provides another illustration of how symmetry can result in degeneracy. If, for example, $K_1 = K_2 = K_3$ it follows from (3.18) that the potential is spherically symmetric and from (3.20) that all states with the same value of $(n_1 + n_2 + n_3)$ are degenerate.

3.3 SEPARATION IN SPHERICAL POLAR COORDINATES

Although there are applications of quantum mechanics in which Cartesian coordinates can be usefully employed, many interesting physical systems, particularly atoms and nuclei, are much more nearly spherical than they are rectangular. Spherically symmetric systems where the potential $V(r)$ is independent of the direction of $\mathbf{r}$, are usually best treated using spherical polar coordinates (r, θ, ϕ). These are related to the Cartesian coordinates (x, y, z) by the expressions

$$\left.\begin{array}{l} x = r \sin\theta \cos\phi \\ y = r \sin\theta \sin\phi \\ z = r \cos\theta \end{array}\right\} \tag{3.22}$$

and the geometrical relationship between the two systems is shown in Fig. 3.2. The Schrödinger equation (3.7) can be written in spherical polar coordinates as

$$-\frac{\hbar^2}{2\mu}\left[\frac{1}{r^2}\frac{\partial}{\partial r}\left(r^2\frac{\partial u}{\partial r}\right) + \frac{1}{r^2 \sin\theta}\frac{\partial}{\partial\theta}\left(\sin\theta\frac{\partial u}{\partial\theta}\right) + \frac{1}{r^2 \sin^2\theta}\frac{\partial^2 u}{\partial\phi^2}\right] + V(r)u = Eu \tag{3.23}$$

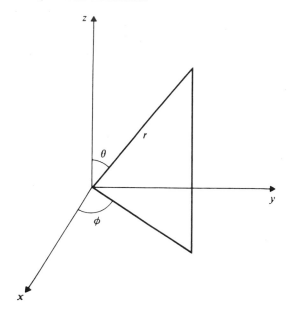

Figure 3.2 The geometrical relationship between the spherical polar coordinates r, θ, ϕ and the Cartesian axes x, y, z.

where the standard expression (derived in many mathematics textbooks) is used to express $\nabla^2 u$ in spherical polar coordinates, and we have represented the particle mass by μ rather than m because the latter symbol will be used later to represent a quantum number.

We now proceed to separate the variables and do so in two stages. We first put $u(r, \theta, \phi) = R(r) Y(\theta, \phi)$, substitute into (3.23), divide through by u and multiply through by r^2 to get

$$
\left[-\frac{\hbar^2}{2\mu} \frac{1}{R} \frac{d}{dr} \left(r^2 \frac{dR}{dr} \right) + r^2 V - r^2 E \right]
$$

$$
+ \left[-\frac{\hbar^2}{2\mu} \frac{1}{Y} \frac{1}{\sin \theta} \frac{\partial}{\partial \theta} \left(\sin \theta \frac{\partial Y}{\partial \theta} \right) - \frac{\hbar^2}{2\mu} \frac{1}{Y} \frac{1}{\sin^2 \theta} \frac{\partial^2 Y}{\partial \phi^2} \right] = 0 \qquad (3.24)
$$

The contents of the first square bracket are independent of θ and ϕ and those of the second are independent of r so they must be separately equal to constants, and the sum of the two constants must be equal to zero. We call these constants $-\lambda$ and λ, and so obtain

$$
-\frac{\hbar^2}{2\mu} \frac{1}{r^2} \frac{d}{dr} \left(r^2 \frac{dR}{dr} \right) + \left(V + \frac{\lambda}{r^2} \right) R = ER \qquad (3.25)
$$

and

$$
-\frac{\hbar^2}{2\mu} \frac{1}{\sin \theta} \frac{\partial}{\partial \theta} \left(\sin \theta \frac{\partial Y}{\partial \theta} \right) - \frac{\hbar^2}{2\mu} \frac{1}{\sin^2 \theta} \frac{\partial^2 Y}{\partial \phi^2} = \lambda Y \qquad (3.26)
$$

Equation (3.26) does not contain the potential V. This means that if we can solve (3.26) for $Y(\theta, \phi)$, the solutions will represent the angular parts of the wave functions for *any* spherically symmetric potential $V(r)$ and we then 'only' have to solve the radial equation (3.25) to get the complete wave function in a particular case. We shall now show how the general solutions to (3.26) are obtained and return to the solution of the radial equation for particular potentials later.

Continuing the separation process, we put $Y(\theta, \phi) = \Theta(\theta)\Phi(\phi)$, substitute into (3.18), divide through by Y, and multiply through by $\sin^2 \theta$ to get

$$\left[-\frac{\hbar^2}{2\mu} \frac{\sin \theta}{\Theta} \frac{d}{d\theta}\left(\sin \theta \frac{d\Theta}{d\theta} \right) - \lambda \sin^2 \theta \right] + \left[-\frac{\hbar^2}{2\mu} \frac{1}{\Phi} \frac{d^2\Phi}{d\phi^2} \right] = 0 \quad (3.27)$$

The contents of the first square bracket are independent of ϕ while those of the second are independent of θ, so they must each be equal to a constant and the sum of the two constants must be equal to zero. We call these constants $-v$ and v, and get

$$-\frac{\hbar^2}{2\mu} \sin \theta \frac{d}{d\theta}\left(\sin \theta \frac{d\Theta}{d\theta} \right) - \lambda \sin^2 \theta \Theta + v\Theta = 0 \quad (3.28)$$

and

$$-\frac{\hbar^2}{2\mu} \frac{1}{\Phi} \frac{d^2\Phi}{d\phi^2} = v \quad (3.29)$$

The solution to (3.29) is straightforward, giving

$$\Phi = A \exp[\pm i(2\mu v/\hbar^2)^{1/2}\phi] \quad (3.30)$$

where A is a constant. We can now apply the condition that the wave function, and hence Φ, must be single valued† so that

$$\Phi(\phi + 2\pi) = \Phi(\phi)$$

Thus

$$\exp[\pm i(2\mu v/\hbar^2)^{1/2}2\pi] = 1$$

and so

$$(2\mu v/\hbar^2)^{1/2} = m \quad (3.31)$$

where m is an integer which can be positive or negative or zero. Substituting back into (3.30), we get

$$\Phi = (2\pi)^{-1/2} \exp(im\phi) \quad (3.32)$$

where the factor $(2\pi)^{-1/2}$ is included as a first step to normalizing the wave

† See footnote to p. 21.

function; it ensures that

$$\int_0^{2\pi} |\Phi|^2 \, d\phi = 1 \tag{3.33}$$

We have now completed the solution of one of the three differential equations and obtained one quantum condition (3.31).

Returning to the equation for Θ (3.28), we can substitute from (3.31) and rearrange to get

$$\sin\theta \frac{d}{d\theta}\left(\sin\theta \frac{d\Theta}{d\theta}\right) + (\lambda' \sin^2\theta - m^2)\Theta = 0 \tag{3.34}$$

where $\lambda' = 2\mu\lambda/\hbar^2$. The solution of this equation is made simpler if we make the substitution $v = \cos\theta$ and write $P(v) \equiv \Theta(\theta)$, leading to

$$\frac{d}{d\theta} = -\sin\theta \frac{d}{dv} = -(1 - v^2)^{1/2} \frac{d}{dv}.$$

Equation (3.34) then becomes

$$\frac{d}{dv}\left[(1 + v^2)\frac{dP}{dv}\right] + \left[\lambda' - \frac{m^2}{1 - v^2}\right]P = 0 \tag{3.35}$$

We first consider the simpler special case where m is equal to zero; Eq. (3.35) is then

$$\frac{d}{dv}\left[(1 - v^2)\frac{dP}{dv}\right] + \lambda'P = 0 \tag{3.36}$$

The method of series solution which was previously employed in the case of the one-dimensional simple harmonic oscillator (Sec. 2.6) can now be applied and we put

$$P = \sum_{p=0}^{\infty} a_p v^p \tag{3.37}$$

Hence

$$\frac{d}{dv}\left[(1 - v^2)\frac{dP}{dv}\right] = \frac{d}{dv}\sum_{p=0}^{\infty}[a_p p v^{p-1} - a_p p v^{p+1}]$$

$$= \sum_{p=0}^{\infty} a_p p(p - 1)v^{p-2} - \sum_{p=0}^{\infty} a_p p(p + 1)v^p$$

$$= \sum_{p=0}^{\infty} [a_{p+2}(p + 2)(p + 1) - a_p p(p + 1)]v^p \tag{3.38}$$

We can now substitute from (3.38) into (3.36):

$$\sum_{p=0}^{\infty} \{a_{p+2}(p+2)(p+1) - a_p[p(p+1) - \lambda']\} v^p = 0$$

This can be true only if the coefficient of each power of v is zero, so we obtain the recurrence relation

$$\frac{a_{p+2}}{a_p} = \frac{p(p+1) - \lambda'}{(p+1)(p+2)}$$

$$\rightarrow 1 \qquad \text{as } p \rightarrow \infty \tag{3.39}$$

Thus, for large p the series (3.37) is identical to the Taylor expansion of the function $(1-v)^{-1}$ which diverges to infinity at the points $v = \pm 1$. Such a divergence in the wave function is not consistent with physical boundary conditions so the series must terminate at some finite value of p, say $p = l$, and we therefore obtain the second quantum condition

$$\lambda' = l(l+1) \tag{3.40}$$

where l is an integer which is greater than or equal to zero, along with the requirement that $a_0 = 0$ if l is odd and $a_1 = 0$ if l is even. Thus $P (\equiv P_l)$ is a polynomial of degree l which contains either only odd powers or only even powers of v. These polynomials are known as the Legendre polynomials and their properties are described in many mathematics textbooks. Explicit forms, corresponding to particular values of l, are easily obtained from (3.40) and (3.39); for example

$$\left.\begin{array}{l} P_0(v) = 1 \\ P_1(v) = v \\ P_2(v) = \frac{1}{2}(3v^2 - 1) \\ P_3(v) = \frac{1}{2}(5v^3 - 3v) \end{array}\right\} \tag{3.41}$$

where the values of the constants a_0 and a_1 have been chosen in accordance with established convention.

The solution of (3.35) in the general case of non-zero values of m is more complicated and the reader is referred to a mathematics textbook for the details. We note that (3.35) is independent of the sign of m, so we expect the solutions to be characterized by l and $|m|$ and we write them as $P_l^{|m|}(v)$. It can be shown† that

$$P_l^{|m|}(v) = (1 - v^2)^{|m|/2} \frac{d^{|m|}P_l}{dv^{|m|}} \tag{3.42}$$

---

†The mathematically inclined reader can verify this result by substituting it into (3.35) and using Leibniz's expression for the nth derivative of a product to show that the result is equivalent to $(1 - v^2)|m|^2$ times the $|m|$th derivative of the left-hand side of Eq. (3.36).

We can use (3.42) to obtain a condition restricting the allowed values of m. P_l is a polynomial of degree l so its $|m|$th derivative, and hence $P_l^{|m|}$, will be zero if $|m|$ is greater than l. But if $P_l^{|m|}$ is zero, the whole wave function must be zero over all space, and this is physically unrealistic. We therefore have the condition

$$-l \leqslant m \leqslant l \tag{3.43}$$

We have now solved the differential equations in θ and ϕ so we can combine the solutions to obtain expressions for the angular part of the wave function, which we now write as $Y_{lm}(\theta, \phi)$, the suffixes l and m emphasizing the importance of these quantum numbers in characterizing the functions.

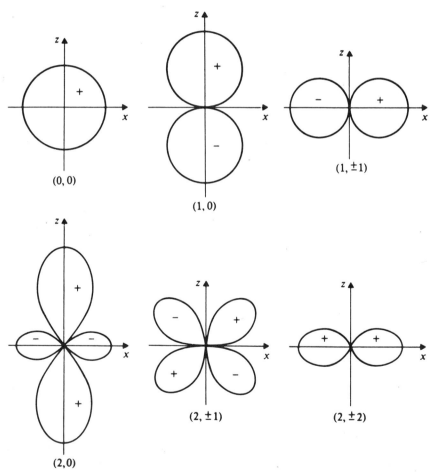

Figure 3.3 Polar plots of the sections at $y = 0$ through the spherical harmonics with quantum numbers (l, m). The distance from the origin of a point on a curve is proportional to the magnitude of the function in that direction. The sign of the function in each region of space is also indicated.

We have

$$Y_{lm}(\theta, \phi) = (-1)^m \left[\frac{(2l+1)}{4\pi} \frac{(l-|m|)!}{(l+|m|)!} \right]^{1/2} P_l^{|m|}(\cos \theta) e^{im\phi} \qquad m \geqslant 0$$

$$(3.44)$$

and $$Y_{l=m}(\theta, \phi) = (-1)^m Y_{lm}(\theta, \phi)$$

where it can be shown that the factor in square brackets ensures normalization of the function when it is integrated over all solid angles; that is,

$$\int_0^{2\pi} \int_0^{\pi} |Y_{lm}(\theta, \phi)|^2 \sin \phi \, d\theta \, d\phi = 1 \qquad (3.45)$$

The phase factors $(-1)^m$ in (3.44) are arbitrary, but chosen in accordance with established convention. The functions Y_{lm} are known as *spherical harmonics* and the reader is once again referred to an appropriate mathematics textbook for a discussion of their properties and a derivation of the form of the normalizing constant. Explicit expressions for the spherical harmonics with l less than or equal to 2 are given below and illustrated in Fig. 3.3 by polar diagrams.

$$Y_{00} = \frac{1}{(4\pi)^{1/2}}$$

$$Y_{10} = \left(\frac{3}{4\pi} \right)^{1/2} \cos \theta$$

$$Y_{1\pm 1} = \mp \left(\frac{3}{8\pi} \right)^{1/2} \sin \theta \, e^{\pm i\phi}$$

$$Y_{20} = \left(\frac{5}{16\pi} \right)^{1/2} (3 \cos^2 \theta - 1) \qquad (3.46)$$

$$Y_{2\pm 1} = \mp \left(\frac{15}{8\pi} \right)^{1/2} \cos \theta \sin \theta \, e^{\pm i\phi}$$

$$Y_{2\pm 2} = \left(\frac{15}{32\pi} \right)^{1/2} \sin^2 \theta \, e^{\pm 2i\phi}$$

A notable feature of Fig. 3.3 is that the wave functions have a particular orientation in space even though the potential is spherically symmetric and the direction of the z axis (sometimes known as the axis of quantization) is therefore arbitrary. This apparent paradox is resolved in the same way as in the similar case of a particle in a square box discussed above. We first note that m does not enter Eq. (3.25) which determines the energy levels of the system, so there are always $2l + 1$ degenerate states that differ only in their values of m. If we measure the energy of such a system, we shall not be able to tell which of these wave functions is appropriate and we must therefore average their squared moduli in order to calculate the position probability distribution. The apparently angularly dependent part of this quantity will therefore be given by

$$(2l+1)^{-1} \sum_{m=-l}^{l} |Y_{lm}(\theta, \phi)|^2$$

It is one of the standard properties of the spherical harmonics that the above quantity is spherically symmetric (as can be readily verified in the cases where $l = 0$, 1 and 2 by substituting the expressions given in Eq. (3.46)) so we once again see that the predictions of quantum mechanics concerning physically measurable quantities are consistent with what would be expected from the symmetry of the problem.

The physical significance of the quantum numbers l and m will be discussed in detail later (Chapter 5). For the moment we note that these cannot be directly connected with the quantization of the energy of the system as the latter quantity appears only in the radial equation which we have yet to solve. It will turn out that l and m are associated with the quantization of the angular momentum of a particle in a central field: the square of the angular momentum has the value $l(l + 1)\hbar^2$ and the z component of angular momentum has the value $m\hbar$.

The Radial Equation

We now turn our attention to the radial equation (3.25) which determines the energy levels of the system. Substituting the expression for λ obtained from the angular solution (3.40) we get

$$-\frac{\hbar^2}{2\mu} \frac{1}{r^2} \frac{d}{dr}\left(r^2 \frac{dR}{dr}\right) + \left[V(r) + \frac{l(l+1)\hbar^2}{2\mu r^2}\right] R = ER$$

This can be simplified by making the substitution $\chi(r) = rR(r)$ which gives

$$-\frac{\hbar^2}{2\mu} \frac{d^2\chi}{dr^2} + \left[V(r) + \frac{l(l+1)\hbar^2}{2\mu r^2}\right] \chi = E\chi \qquad (3.47)$$

Apart from the second term within the square brackets, Eq. (3.47) is identical in form to the one-dimensional Schrödinger equation. However, an additional boundary condition applies in this case: χ must equal zero at $r = 0$ otherwise $R = r^{-1}\chi$ would be infinite at that point.†

As well as being mathematically convenient, the function $\chi(r)$ has a physical interpretation in that $|\chi|^2 dr$ is the probability of finding the electron

---

†The observant reader will have noticed that if $R \sim r$ then $\int_0^r |R|^2 r^2 dr$ will be finite and therefore the boundary condition set out in Chapter 2 is not breached. It can be shown that the condition $\chi = 0$ at $r = 0$ follows from the requirement that the solutions of the Schrödinger equation expressed in spherical polar coordinates must also be solutions when the equation is written in Cartesians. Further details on this point can be found in P. A. M. Dirac *The Principles of Quantum Mechanics* (Oxford 1974) Chapter 6.

at a distance between r and $r + dr$ from the origin averaged over all directions. This follows from the fact that this probability is obtained by integrating $|\psi(r, \theta, \phi)|^2$ over a spherical shell of radius r and thickness dr. That is, it is given by

$$|R^2(r)|r^2\, dr \int_0^{2\pi} \int_0^{\pi} |Y(\theta, \phi)^2|\sin\theta\, d\theta\, d\phi$$

$$= |\chi^2(r)|$$

using (3.45).

To progress further with the solution of the radial equation, the form of the potential $V(r)$ must be known, and in the next section we shall consider the particular example of the hydrogenic atom.

3.4 THE HYDROGENIC ATOM

We are now ready to apply quantum theory to the real physical situation of an electron moving under the influence of a positively charged nucleus. If this nucleus consists of a single proton, the system is a hydrogen atom, but the theory is also applicable to the more general case of an atom with atomic number Z (and hence nuclear charge Ze) with all but one of its electrons removed (for example, He^+, Li^{++}, etc.); in general such a system is described as a *hydrogenic atom*. The potential energy of interaction between the electron and the nucleus is $-Ze^2/4\pi\varepsilon_0 r$, so Eq. (3.47) becomes† in this case

$$-\frac{\hbar^2}{2\mu}\frac{d^2\chi}{dr^2} + \left[-\frac{Ze^2}{4\pi\varepsilon_0 r} + \frac{l(l+1)\hbar^2}{2\mu r^2}\right]\chi = E\chi \qquad (3.48)$$

The solution of Eq. (3.48) will again involve considerable manipulation which is simplified by making a suitable substitution. We define a new variable ρ so that

$$\rho = (-8\mu E/\hbar^2)^{1/2}r \qquad (3.49)$$

(note that E is negative for bound states as the potential is zero when r is

---

†We have assumed above that the electron is moving in the field of a fixed nucleus, but this will not be exactly true as the nucleus is also moving in the field of the electron. As is shown in Chapter 10, this nuclear motion can be allowed for in exactly the same way as in classical mechanics by taking r to be the distance between the nucleus and the electron, and μ to be the reduced mass of the nucleus (mass m_N) and the electron (mass m_e). That is,

$$\mu = m_N m_e/(m_N + m_e)$$

Because the mass of the electron is much smaller than that of the nucleus, μ is very nearly equal to m_e and the effect of nuclear motion is small, although large enough to have a significant effect on the comparison between the theoretical and experimental values of the Rydberg constant discussed below.

infinite) and hence

$$\frac{d^2\chi}{dr^2} = -\frac{8\mu E}{\hbar^2}\frac{d^2\chi}{d\rho^2} \tag{3.50}$$

Equation (3.48) now becomes

$$\frac{d^2\chi}{d\rho^2} - l(l+1)\frac{\chi}{\rho^2} + \left(\frac{\beta}{\rho} - \frac{1}{4}\right)\chi = 0 \tag{3.51}$$

where the constant β is defined as

$$\beta = \left(-\frac{\mu}{2E}\right)^{1/2}\frac{Ze^2}{4\pi\varepsilon_0\hbar} \tag{3.52}$$

We first consider the solution to (3.51) in the case of very large ρ when the equation becomes

$$\frac{d^2\chi}{d\rho^2} - \tfrac{1}{4}\chi = 0 \tag{3.53}$$

leading to

$$\chi \sim \exp(-\rho/2) \tag{3.54}$$

(where we have rejected a possible solution with positive exponent because it diverges to infinity at large ρ). This suggests that we try

$$\chi = F(\rho)\exp(-\rho/2) \tag{3.55}$$

as a solution to (3.51). On substitution we get

$$\frac{d^2F}{d\rho^2} - \frac{dF}{d\rho} - \frac{l(l+1)}{\rho^2}F + \frac{\beta}{\rho}F = 0 \tag{3.56}$$

We now look for a series solution to (3.56) and put

$$F = \sum_{p=1}^{\infty} a_p\rho^p \tag{3.57}$$

The lower limit of this summation is $p = 1$ rather than $p = 0$, otherwise F and, therefore, χ would not be zero at $\rho = 0$. Thus

$$\frac{dF}{d\rho} = \sum_{p=1}^{\infty} pa_p\rho^{p-1} \tag{3.58}$$

and

$$\frac{d^2F}{d\rho^2} = \sum_{p=1}^{\infty} p(p-1)a_p\rho^{p-2}$$

$$= \sum_{p=1}^{\infty} (p+1)pa_{p+1}\rho^{p-1} \tag{3.59}$$

Also

$$F/\rho^2 = \sum_{p=1}^{\infty} a_p \rho^{p-2}$$

$$= a_1 \rho^{-1} + \sum_{p=1}^{\infty} a_{p+1} \rho^{p-1} \tag{3.60}$$

Substituting from Eqs (3.57) to (3.60) into (3.56) we get

$$-l(l+1)a_1\rho^{-1} + \sum_{p=1}^{\infty} [(p+1)pa_{p+1} - pa_p - l(l+1)a_{p+1} + \beta a_p]\rho^{p-1}$$

$$\tag{3.61}$$

The coefficient of each power of ρ must vanish so we have

$$a_1 = 0 \qquad \text{unless } l = 0$$

and

$$\frac{u_{p+1}}{a_p} = \frac{p - \beta}{p(p+1) - l(l+1)} \tag{3.62}$$

$$\rightarrow p^{-1} \qquad \text{as } p \rightarrow \infty \tag{3.63}$$

We first note that the denominator on the right-hand side of (3.62) is zero if $p = l$. This implies that a_{l+1} (and, by implication, all other a_p where p is greater than l) must be infinite unless a_l is zero. But if a_l equals zero then it also follows from (3.62) that a_{l-1}, a_{l-2} etc must also equal zero. We conclude, therefore, that all a_p with p less than or equal to l must be zero if the solution is to represent a physically realistic wave function. We also see that (3.63) is identical to the recurrence relation for the terms in the series expansion of $\exp(\rho)$ and so χ, which equals $F\exp(-\rho/2)$, will diverge like $\exp(\rho/2)$ as ρ tends to infinity. However, just as in the case of solutions to the harmonic oscillator and Legendre polynomial equations, this divergence can be prevented by ensuring that the series terminates after a finite number of terms. For this to occur at the term $p = n$ we must have

$$\beta = n > l \tag{3.64}$$

and hence, using (3.52)

$$E \equiv E_n = -\frac{\mu Z^2 e^4}{2(4\pi\varepsilon_0)^2 \hbar^2 n^2} \tag{3.65}$$

We have thus derived expressions for the discrete energy levels of the hydrogenic atom in terms of the reduced mass of the electron and nucleus, the nuclear charge, and the fundamental constants e, $\hbar$, and ε_0. It should be noted that the energy levels (3.65) are not only independent of m, as would be expected from the earlier discussion, but are also independent of l. This additional degeneracy is a particular feature of the Coulomb potential and is not a general property of a spherically symmetric system.

It is now an acid test of the theory developed so far that we compare the above energy levels with those experimentally measured from observations of atomic spectra. We saw in Chapter 1 that the line spectra of hydrogen could be accounted for if the hydrogen atom were assumed to have a set of energy levels given by

$$E_n = -2\pi\hbar c R_0/n^2 \tag{3.66}$$

where n is a positive integer and R_0 is a constant whose currently accepted best value is $1.096\,775\,9\,(1) \times 10^7$ m^{-1}, the bracketed number indicating the experimental error in the last place. Comparison of (3.65) and (3.66) shows at once that these have the same form so that there is at least qualitative agreement between theory and experiment. Quantitative comparison is made using the measured values of the fundamental constants

$$\mu = 9.104\,575\,(89) \times 10^{-31}\ \text{kg}$$

$$\varepsilon_o = 8.854\,187\,82\,(7) \times 10^{-12}\ \text{F m}^{-1}$$

$$\hbar = 1.054\,588\,7\,(57) \times 10^{-34}\ \text{J s}$$

$$e = 1.602\,189\,2\,(46) \times 10^{-19}\ \text{C}$$

$$c = 2.997\,924\,6\,(1) \times 10^8\ \text{m s}^{-1}$$

to obtain an estimate of R_0 from Eq. (3.65) as $1.096\,775\,7\,(95) \times 10^7$ m^{-1}. Thus the agreement between theory and experiment is well within the range of the extremely small experimental errors. Similar agreement is obtained for other hydrogenic atoms when the appropriate values of the nuclear charge and the reduced mass are substituted into Eq. (3.65). These results therefore represent an important test of quantum mechanical theory which it has passed with flying colours.† Our belief in quantum mechanics does not of course rest on this result alone, and indeed an identical expression to (3.65) was derived by Niels Bohr using an earlier theory which was subsequently shown to be incorrect when applied to other more complex systems. However, although we shall compare the results of calculation and experiment on a number of other occasions when we shall always find agreement within the limits of experimental error, there are very few examples of physical quantities whose values can be both measured experimentally to such high precision and also calculated exactly by solving the appropriate quantum-mechanical equations.

---

†Of course the quantum theory of atomic spectra is now so well established that formulae such as (3.65) are themselves used in determining the best values of the fundamental constants, but the fact that a wide variety of experimental data can be successfully and consistently used in this way is itself a confirmation of the theory.

The Hydrogenic Atom Wave Functions

We now complete our consideration of the hydrogenic atom by discussing the form of the wave functions associated with the different energy levels. We saw above that the radial part of the wave function is consistent with the boundary conditions only if the series (3.57) for F starts at the term $p = l + 1$ and terminates at $p = n$. We thus have

$$F_n(\rho) = \sum_{p=l+1}^{n} a_p \rho^p \tag{3.67}$$

where the coefficients a_p can be expressed in terms of a_{l+1} using the recurrence relation (3.62) with $\beta = n$. The polynomials so obtained are known as the *associated Laguerre functions*. We can then use (3.55) and the definition of ρ in terms of r to obtain $\chi(r)$ and hence $R(r)$. This can be combined with the appropriate spherical harmonic to produce an expression for the complete time-independent part of the wave function, $u(r, \theta, \phi)$, which will be normalized, provided the spherical harmonic has been normalized in accordance with (3.45) and the constant a_{l+1} has been chosen so that

$$\int_0^\infty |R|^2 r^2 \, dr = 1 \tag{3.68}$$

Formally, then, we have

$$u_{nlm} = R_{nl}(r) Y_{lm}(\theta, \phi) \tag{3.69}$$

where the suffixes indicate the dependence of the various functions on the quantum numbers n, l, and m. The wave functions corresponding to the five states of lowest energy as determined in this way are

$$\left.\begin{aligned}
u_{100} &= (Z^3/\pi a_0^3)^{1/2} \exp(-Zr/a_0) \\
u_{200} &= (Z^3/8\pi a_0^3)^{1/2} (1 - Zr/2a_0) \exp(-Zr/2a_0) \\
u_{210} &= (Z^3/32\pi a_0^3)^{1/2} (Zr/a_0) \cos\theta \exp(-Zr/2a_0) \\
u_{21\pm1} &= \mp (Z^3/\pi a_0^3)^{1/2} (Zr/8a_0) \sin\theta \exp(\pm i\phi) \exp(-Zr/2a_0)
\end{aligned}\right\} \tag{3.70}$$

where the constant a_0 is defined as

$$a_0 = 4\pi\varepsilon_o \hbar^2/\mu e^2$$
$$= 0.529\,177 \times 10^{-10} \text{ m} \tag{3.71}$$

and is known as the *Bohr radius*.

The value of the azimuthal quantum number l is often denoted by a particular letter code: states with $l = 0$, 1, 2, and 3 are labelled s, p, d, and f respectively. This letter is sometimes prefixed by a number equal to the quantum number n; thus for example, the first state in (3.70) is known as the 1s state, the second is 2s, and the others are 2p states.

The radial parts of the wave functions (3.70) are plotted as functions of

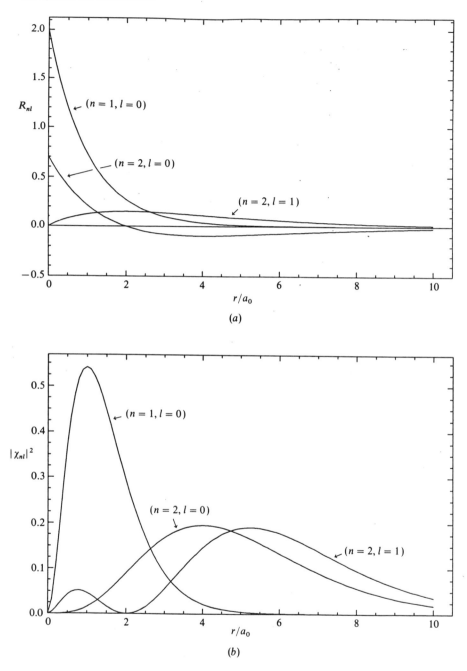

Figure 3.4 The radial parts, $R_{n,l}$, of the wave functions corresponding to some of the energy states of the hydrogen atom are shown in (a). The corresponding radial probability distributions, $|\chi_{n,l}|^2 = r^2 |R_{n,l}|^2$, are displayed in ($b$).

r in Fig. 3.4 for the case of the hydrogen atom where $Z = 1$. We see that the constant a_0 characterizes the width of the wave function of the lowest energy state and that this width increases for states of higher energy. We can combine (3.71) and (3.65) to express the energy levels in terms of a_0:

$$E_n = -\frac{Z^2 e^2}{2(4\pi\varepsilon_0)a_0 n^2} \tag{3.72}$$

As the potential energy is given by $V = -Ze^2/4\pi\varepsilon_0 r$, an electron with the above total energy could only have positive kinetic energy for values of r less than $2n^2 a_0/Z$. These limits are indicated in Fig. 3.4 and we see that the exponential tails of the wave functions penetrate the classically forbidden region in a manner very similar to that discussed in the one-dimensional cases in Chapter 2.

Figure 3.4 also shows $|\chi^2(r)| = r^2|R^2(r)|$ for each state as a function of r. As we pointed out earlier, this expression equals the probability that the electron be found at a distance between r and $r + dr$ from the origin (in any direction). We see that this probability reaches a maximum at $r = a_0$ in the case of the ground state wave function. We particularly note that in all cases $|\chi^2|$ equals zero at the origin, even though the square of the wave function is actually a maximum at that point. The reader should think carefully about this apparent contradiction and how it can be resolved by understanding the different nature of the two probability distributions represented by $|\psi^2|$ and $|\chi^2|$.

PROBLEMS

3.1 Calculate the energy levels and obtain expressions for the associated wave functions in the case of a particle moving in two dimensions in a rectangular, infinite-walled box of sides a and b. Discuss the degeneracy of the system and the symmetry of the position probability distribution when $a = b$.

3.2 What is the symmetry of the position probability distribution and how is it related to the degeneracy in the case of a particle confined to a three-dimensional box with cubic symmetry (that is, with $a = b = c$)?

3.3 A particle moves in two dimensions in a circularly symmetric potential. Show that the time-independent Schrödinger equation can be separated in plane polar coordinates and that the angular part of the wave function has the form $(2\pi)^{1/2} \exp(im\phi)$ where m is an integer. What is the symmetry of the position probability distribution in this case?

3.4 Consider a circularly symmetric two-dimensional system similar to that described in Prob. 3.3 where the potential is zero for all values of r less than a and infinite otherwise. Show that the radial part $R(r)$ of the wave function must satisfy the equation

$$\frac{d^2 R}{d\rho^2} + \frac{1}{\rho}\frac{dR}{d\rho} + \left(1 - \frac{m^2}{\rho^2}\right)R = 0$$

where $\rho = (2\mu E)^{1/2} r$. In the case where $m = 0$ show that $R = \sum_{k=0}^{\infty} A_k \rho^k$ where $A_k = 0$ if k is odd and $A_{k+2} = -A_k/(k+2)^2$. Given that the first zero of this function is at $\rho = 2.405$, obtain an expression for the energy of the ground state of the system.

3.5 A particle of mass μ moves in a three-dimensional spherically symmetric well where $V = 0$, $r \leqslant a$ and $V = V_0, r > a$. Show that the energies of those states with quantum number $l = 0$ are determined by the condition $k \cot ka = -\kappa$ in the notation used in Chapter 2. Show that there are no bound states of such a system unless $V_0 > \hbar^2 \pi^2 / 8ma^2$. Would you expect this condition to be modified if states with $l \neq 0$ were also considered? Does a similar condition hold in the case of a cubic well?

3.6 The (negative) binding energy of the ground state of the deuteron (neutron + proton) is 2.23 MeV. Assuming that the interaction potential is of the form described in Prob. 3.5 with $a = 2.0 \times 10^{-15}$ m, find the corresponding value of V_0. Do bound states of the deuteron other than the ground state exist?

   Hints: Remember to use the reduced mass; $x = 1.82$ is a solution to the equation $x \cot x = -0.46$.

3.7 Verify that

$$\int_0^{2\pi} \int_0^{\pi} Y^*_{lm} Y_{l'm'} \sin \theta \, d\theta \, d\phi = 0 \qquad \text{unless } l = l' \text{ and } m = m'$$

for all values of l, l', m and m' up to and including those with l and/or l' equal to 2, where Y_{lm} is a spherical harmonic and Y^*_{lm} its complex conjugate.

3.8 Use the hydrogen atom wave functions and the probabilistic interpretation of the wave function to calculate (i) the most probable and (ii) the average values of the distance between the electron and proton in a hydrogen atom in its 1s state.

FOUR

THE BASIC POSTULATES OF QUANTUM MECHANICS

In the previous chapters we have seen how solutions to the time-independent Schrödinger equation correspond to the allowed energy levels of a quantum system and how, in the hydrogen atom case in particular, the results of this procedure are in extremely good agreement with experiment. It would be possible at this stage to extend the process to a prediction of the energy levels of other atoms. We would find that the corresponding Schrödinger equations could no longer be solved exactly, but that approximations could be developed which, when combined with computational techniques, would lead to predicted energy levels that were once again in very good agreement with experiment. However, such a programme, most of which is beyond the scope of this book in any case, would be premature at this stage as we have not yet established a general procedure which will do more than predict the allowed energy levels of a particle moving in a potential and the probability that it is in the vicinity of a particular point in space. We do not yet know, for example, how to predict the momentum of an electron in a hydrogen atom; we do not even know if it has a definite value or if it can only be specified by a probability distribution as is the case for electron position. We have as yet no way of predicting under what conditions an atom undergoes a transition from one state to another, emitting energy in the form of a quantum of electromagnetic radiation.

Answers to the above and similar questions require a more general and more sophisticated approach to quantum mechanics. We shall develop the foundations of this in the present chapter and we shall see that it is based on five postulates. In developing and discussing the postulates, we shall build on what we have already done in previous chapters to make each postulate seem at least reasonable. It is important to remember that this is an *inductive* process; i.e., although the postulates can be made to appear reasonable by considering particular examples, they can never be rigorously derived in this way. We encountered a previous example of an inductive argument when we set up the Schrödinger equation on a basis of a consideration of classical waves and the experimental de Broglie relations; we emphasized then that this argument in no way represented a proof of the Schrödinger equation and that our belief in its correctness lay in the fact that it was successful in predicting the results of experiments. In a similar way the correctness or otherwise of the basic postulates of quantum mechanics rests on the agreement between deductions from them and the results of experiments. To the present time quantum mechanics has withstood every experimental test; there are many fields of physics that are not yet completely understood, but there are no experimental results, even in the field of particle physics, which contradict or falsify the fundamental principles of quantum mechanics. Despite this, the fundamental basis of the subject contains ideas that run very much against the intuition we have all developed in our interaction with the classical world. We have come across some of these already in our discussion of wave–particle duality in earlier chapters and we shall come across some more in this chapter and later. However, we shall postpone any detailed discussion of the foundations of the subject to the last chapter, where we shall attempt to introduce the reader to the main ideas underlying the vigorous philosophical debate about the conceptual basis of quantum mechanics that has gone on ever since the early days of the subject.

It might therefore have been expected that we should simply have stated the basic postulates at the beginning of this book, thereafter concentrating on deriving predictions of experimental results from them, and indeed such an approach is sometimes adopted. However, the ideas of quantum mechanics are rather abstract and most students understand them more easily when introduced to them gradually using inductive arguments based on a knowledge of the wave mechanics of atoms as described in the previous chapters.

4.1 THE WAVE FUNCTION

The experimental evidence for the wave properties of the electron and the success of the Schrödinger equation in predicting the energy levels of the hydrogen atom indicate that the physical properties of a particle moving under the influence of a potential are contained in the *wave function*. As we saw, this is a single-valued function of the coordinates of the particle and the time; it is finite, differentiable, and can be integrated over all space. It is important to remember that, although the measurable properties of a system are derived from the wave function, it is not itself a physical quantity. On the other hand it is a fundamental principle of quantum mechanics that the wave function contains all the information it is possible to obtain about a physical system in a particular state. The first postulate then concerns the existence of the wave function.

> **Postulate 4.1** For every dynamical system there exists a wave function that is a continuous, integrable, single-valued function of the parameters of the system and of time, and from which all possible predictions of the physical properties of the system can be obtained.

The above statement covers the case of a particle moving in a potential where the 'parameters of the system' are the particle coordinates, but it also refers to more general situations where the parameters may be, for example, the coordinates of all the particles of a many-body system and may also include internal variables such as 'spin'.

Notation

In the previous chapters we used the symbol Ψ to represent a general solution to the time-dependent Schrödinger equation and the symbol u (sometimes with a subscript) as the time-independent part of the wave function of a system in a state of given energy. We shall continue this notation in the present chapter, and in addition we shall use the symbol ψ to represent a general wave function, whose time dependence we are not explicitly considering, and the symbol ϕ (often with a subscript) when the system is in what we shall call an 'eigenstate'—that is when some dynamical quantity (not necessarily the energy) has a known value.

4.2 THE DYNAMICAL VARIABLES

In this section we consider how the dynamical variables (position, momentum, angular momentum, energy, etc.) are represented mathematically in quantum mechanics. In classical mechanics, of course, such quantities are represented by algebraic variables: position coordinates x, y, z; momentum components p_x, p_y, p_z; energy E; etc. However, algebraic variables can take on any value whereas we have seen that quantum mechanical properties are often confined to a discrete set of values (e.g., the energy levels of a hydrogen atom). Moreover, the uncertainty principle indicates that simultaneous specification of the magnitudes of two dynamical quantities (e.g., position and momentum) is not always possible in quantum mechanics whereas algebraic variables always imply precise values. We must therefore look for a new way of representing dynamical quantities mathematically.

Let us consider again the time-independent Schrödinger equation in the case of a particle moving in a potential $V(\mathbf{r})$, which was given in Eq. (3.7) and which we now write in the form

$$\left[-\frac{\hbar^2}{2m} \nabla^2 + V(\mathbf{r}) \right] u_n = E_n u_n \qquad (4.1)$$

or

$$\hat{H} u_n = E_n u_n \qquad \text{where } \hat{H} = -\frac{\hbar^2}{2m} \nabla^2 + V \qquad (4.2)$$

The quantity in square brackets in (4.1), which is defined as $\hat{H}$ in (4.2), is an example of a mathematical *operator*. Such operators operate on functions (in this case the u_n) to produce new functions, and Eq. (4.2) therefore states that the energy levels E_n and corresponding wave functions u_n are such that, when the operator $\hat{H}$ operates on u_n, it produces a result equivalent to multiplying u_n by the constant E_n. The quantities E_n and the functions u_n are known as the *eigenvalues* and *eigenfunctions* respectively of the operator $\hat{H}$, and we say that the energy of the quantum-mechanical system is represented by an operator $\hat{H}$ whose eigenvalues are equal to the allowed values of the energy of the system. For historical reasons $\hat{H}$ is often known as the *Hamiltonian operator*.

We now consider the significance of the eigenfunctions u_n. In the earlier chapters we interpreted u_n as representing the wave function of the system when it was in a state whose energy is E_n. This implies that if the energy of the system were to be measured when the wave function is u_n, we should certainly obtain the result E_n. As a second measurement of the energy performed immediately after the first would be reasonably expected to yield the same result, we conclude that the wave function of a quantum mechanical system will be identical to the corresponding eigenfunction of the Hamiltonian operator *immediately after* a measurement of the energy of the system.

It is a reasonable extension of the above arguments to say that other dynamical variables (position, momentum, etc.) should also be capable of representation by operators, that the eigenvalues of these operators should correspond to the possible results of experiments carried out to measure these quantities, and that the wave function of a quantum mechanical system should be identical to the corresponding eigenfunction following the measurement represented by such an operator. Consider first the momentum operator: the first part of the Hamiltonian (4.2) clearly corresponds to the kinetic energy of the particle, and classically this is related to the particle momentum by the expression

$$T = p^2/2m \qquad (4.3)$$

If we assume that a similar relation holds in quantum mechanics we get

$$\frac{1}{2m} \hat{P}^2 = -\frac{\hbar^2}{2m} \nabla^2 \qquad (4.4)$$

where $\hat{P}$ is the momentum operator. An expression for $\hat{P}$ which is consistent with the above is

$$\hat{P} = -i\hbar \nabla$$

that is,

$$\hat{P}_x = -i\hbar \frac{\partial}{\partial x} \qquad \text{etc.}$$

$$(4.5)$$

where $\hat{P}_x$ is the operator representing the x component of the momentum. (The sign chosen on the right-hand side of (4.5) is conventional: a positive sign could equally well be chosen without affecting any of the physical predictions of the theory.)

We now consider the eigenvalue equation of the operator representing the x component of momentum: let the eigenvalues and eigenfunctions be p and ϕ respectively, then

$$-i\hbar \frac{\partial}{\partial x} \phi = p\phi \qquad (4.6)$$

that is,

$$\phi = A \exp(ikx) \qquad (4.7)$$

where A is a constant and $k = p/\hbar$. But this is just the de Broglie relation connecting wave number and momentum (2.1), and Eq. (4.7) is therefore consistent with the discussion in earlier chapters. We now note two points about the momentum eigenvalue equation. Firstly, there is a solution to (4.6) for any value of p so the possible values of the momentum of a particle are not confined to a discrete set in the way that the allowed energy levels of a bound particle are. Secondly, the momentum eigenfunctions are always different from the energy eigenfunctions (except in the case of a free particle when $V(r)$ is constant everywhere). Thus if the energy of a bound particle is measured, the wave function immediately after the measurement will not be an eigenfunction of the momentum operator. We shall return to this point in more detail later when we discuss the fourth postulate, but in the meantime we simply note that the outcome of a measurement of the momentum of a bound particle when the energy is already known cannot be accurately predicted.

We now consider the operator representing the position of a particle, confining our discussion to one dimension for the moment. The eigenfunction equation can be written formally as

$$\hat{X}\phi = x_0\phi \qquad (4.8)$$

where x_0 and ϕ now represent the position eigenvalues and eigenfunctions respectively and we have to obtain a form of $\hat{X}$ which is consistent with what we already know. We remember that in the previous chapters the squared modulus of the wave function was interpreted as the probability density for the position of a particle. If, therefore, the eigenfunction ϕ is to represent the wave function immediately after a position measurement yielding the result x_0, it follows that ϕ must be very large at the point $x = x_0$ and vanishingly small everywhere else. A function with such properties is known as the *Dirac delta function* and is written as $\delta(x - x_0)$. Thus we have

$$\hat{X}\delta(x - x_0) = x_0\delta(x - x_0) \qquad (4.9)$$

which equation is satisfied if

$$\hat{X} \equiv x \qquad (4.10)$$

as can be seen by substituting (4.10) into (4.9) and considering separately the point $x = x_0$ and the region $x \neq x_0$. That is, the quantum mechanical operator representing the x coordinate of a particle is just the algebraic variable x. In three dimensions, it follows that the particle position operator is the vector $\mathbf{r}$ and we note that this is consistent with the fact that we wrote the energy operator in the form

$$\hat{H} = -\frac{\hbar^2}{2m}\nabla^2 + V(\mathbf{r})$$

where $V(\mathbf{r})$ is just the potential expressed as a function of the algebraic variable $\mathbf{r}$. We thus conclude that the position and momentum of a particle can be represented by the operators $\mathbf{r}$ and $-i\hbar\nabla$ respectively. We saw that the kinetic and potential energy operators have the same functional dependence on the operators representing position and momentum as do the corresponding quantities in classical mechanics and we assume that this relationship also holds for other dynamical quantities which can be expressed classically as functions of $\mathbf{r}$ and $\mathbf{p}$. Before expressing all this in the form of formal postulates, however, we shall establish one general property of operators used to represent dynamical variables.

We have interpreted the eigenvalues of the operators discussed so far as representing the possible results of experiments carried out to measure the corresponding physical quantities. If this interpretation is to be correct, it is clear that these eigenvalues must be real numbers even though the eigenfunctions, or the operators themselves, may be imaginary (like the momentum operator) or complex. One class of mathematical operators that always have real eigenvalues consists of the *Hermitian* operators. Hermitian operators are defined such that if $f(\mathbf{r})$ and $g(\mathbf{r})$ are any well-behaved functions of $\mathbf{r}$ which vanish at infinity then the operator $\hat{Q}$ is Hermitian if and only if

$$\int f\hat{Q}g\, d\tau = \int g\hat{Q}^*f\, d\tau \qquad (4.11)$$

where $\hat{Q}^*$ is the complex conjugate of $\hat{Q}$ (obtained by replacing i with $-i$ wherever it appears in $\hat{Q}$)† and the integrals are over all space. We can easily show that the eigenvalues of a Hermitian operator are real:

let
$$\hat{Q}\phi_n = q_n\phi_n$$

then
$$\hat{Q}^*\phi_n^* = q_n^*\phi_n^*$$

Hence
$$\int \phi_n^* \hat{Q}\phi_n \, d\tau = q_n \int \phi_n^*\phi_n \, d\tau \tag{4.12}$$

and
$$\int \phi_n \hat{Q}^*\phi_n^* \, d\tau = q_n^* \int \phi_n\phi_n^* \, d\tau \tag{4.13}$$

But if $\hat{Q}$ is Hermitian it follows from (4.11) (identifying f with ϕ_n and g with ϕ_n^*) that the left-hand sides of (4.12) and (4.13) must be equal. Equating the right-hand sides of these equations we get

$$q_n \int |\psi_n|^2 \, d\iota = q_n^* \int |\phi_n|^2 \, d\tau \tag{4.14}$$

and hence
$$q_n = q_n^* \tag{4.15}$$

so that q_n is real as required.

The converse of this theorem (that all operators with real eigenvalues are Hermitian) is not true: the Hermitian property is a stronger condition on the operator than the reality of the eigenvalues. However, as it has been found possible to represent all physical quantities by Hermitian operators, and as these have a number of useful properties, some of which will be discussed later, it is convenient to impose this Hermitian condition.

We can readily show that the operators representing position and momentum in one dimension are Hermitian: in the case of x this follows trivially from substitution into the one-dimensional equivalent of (4.11) while in the case of $\hat{P}_x$ we have

$$\int_{-\infty}^{\infty} f\hat{P}_x g \, dx = -i\hbar \int_{-\infty}^{\infty} f \frac{\partial g}{\partial x} \, dx$$

Integrating by parts, the right-hand side becomes

$$-i\hbar \left\{ [fg]_{-\infty}^{\infty} - \int_{-\infty}^{\infty} g \frac{\partial f}{\partial x} \, dx \right\}$$

---

† A more rigorous definition of the complex conjugate of an operator is that, for any function f, $\hat{Q}^*f^*$ must equal $(\hat{Q}f)^*$. However, in nearly all cases, this definition is equivalent to that given in the text.

The first term is zero because f and g vanish at infinity so we have

$$\int_{-\infty}^{\infty} f \hat{P}_x g \, dx = i\hbar \int_{-\infty}^{\infty} g \frac{\partial f}{\partial x} \, dx$$

$$= \int_{-\infty}^{\infty} g \hat{P}_x^* f \, dx$$

as required. We leave it as an exercise for the reader to extend this argument to three dimensions and to confirm that other operators considered so far are Hermitian.

We can now summarize the contents of the above paragraphs in two postulates.

Postulate 4.2 Every dynamical variable may be represented by a Hermitian operator whose eigenvalues represent the possible results of carrying out a measurement of the value of the dynamical variable. Immediately after such a measurement, the wave function of the system will be identical with the eigenfunction corresponding to the eigenvalue obtained as a result of the measurement.

Postulate 4.3 The operators representing the position and momentum of a particle are $\mathbf{r}$ and $-i\hbar\nabla$ respectively. Operators representing other dynamical quantities bear the same functional relation to these as do the corresponding classical quantities to the classical position and momentum variables.

Before proceeding to develop further postulates we shall establish some mathematical properties of Hermitian operators that will be useful at a later stage.

4.3 THE PROPERTIES OF HERMITIAN OPERATORS

We first show that if $\hat{Q}$ and $\hat{R}$ are Hermitian operators then, although the product $\hat{Q}\hat{R}$ need not be Hermitian, the expressions $(\hat{Q}\hat{R} + \hat{R}\hat{Q})$ and $i(\hat{Q}\hat{R} - \hat{R}\hat{Q})$ are Hermitian. We have

$$\int f \hat{Q}\hat{R}g \, d\tau = \int (\hat{R}g)\hat{Q}^* f \, d\tau$$

using (4.11) and the fact that $\hat{Q}$ is Hermitian. Thus

$$\int f \hat{Q}\hat{R}g \, d\tau = \int (\hat{Q}^* f) R g \, d\tau$$

$$= \int g \hat{R}^* \hat{Q}^* f \, d\tau \qquad (4.16)$$

using the Hermitian property of R. Similarly

$$\int\int f\hat{R}\hat{Q}g \, d\tau = \int g\hat{Q}^*\hat{R}^*f \, d\tau \tag{4.17}$$

Thus, adding (4.16) and (4.17)

$$\int\int f(\hat{Q}\hat{R} + \hat{R}\hat{Q})g \, d\tau = \int g(\hat{Q}\hat{R} + \hat{R}\hat{Q})^*f \, d\tau \tag{4.18}$$

and, subtracting (4.16) and (4.17) and multiplying by i

$$\int\int f[i(\hat{Q}\hat{R} - \hat{R}\hat{Q})]g \, d\tau = \int g[i(\hat{Q}\hat{R} - \hat{R}\hat{Q})]^*f \, d\tau \tag{4.19}$$

which proves the above statements. These results will be used later in this chapter when discussing the uncertainty principle. In the meantime we note that it is an obvious corollary of the above that if $\hat{Q}$ is Hermitian so is $\hat{Q}^2$. Thus, for example, the Hermitian property of the kinetic energy operator, $\hat{T} = \hat{P}^2/2m$, follows directly from the fact that $\hat{P}$ is Hermitian.

Orthonormality

One important property of the eigenfunctions of a Hermitian operator is known as *orthonormality* and is expressed by the following relation:

$$\int \phi_n^* \phi_m \, d\tau = \delta_{nm} \tag{4.20}$$

where ϕ_n and ϕ_m are eigenfunctions of some Hermitian operator and δ_{nm} is the *Kronecker delta* defined by $\delta_{nm} = 0, n \neq m; \delta_{nm} = 1$. To prove this consider a Hermitian operator $\hat{Q}$ whose eigenvalues are q_n and whose eigenfunctions are ϕ_n. That is,

$$\hat{Q}\phi_n = q_n\phi_n \tag{4.21}$$

Thus

$$\int \phi_n^* \hat{Q}\phi_m \, d\tau = q_m \int \phi_n^* \phi_m \, d\tau \tag{4.22}$$

(because q_m is a constant which can be taken outside the integral) and similarly (remembering that q_n is real)

$$\int \phi_m \hat{Q}^* \phi_n^* \, d\tau = q_n \int \phi_m \phi_n^* \, d\tau \tag{4.23}$$

But it follows from the definition of the Hermitian operator (4.11) that the left-hand sides of (4.22) and (4.23) are equal. We thus have, equating the right-hand sides,

$$\int \phi_n^* \phi_m \, d\tau = 0 \quad \text{or} \quad q_m = q_n \tag{4.24}$$

The second alternative implies that $m = n$ or that the eigenfunctions ϕ_m and ϕ_n correspond to the same eigenvalue—i.e., that they are degenerate. We postpone a discussion of degeneracy until later in this chapter and conclude for the moment that

$$\int \phi_n^* \phi_m \, d\tau = 0 \quad m \neq n \tag{4.25}$$

Turning now to the case $m = n$, it is clear from the eigenvalue equation (4.21) that if ϕ_n is an eigenfunction of $\hat{Q}$ then so is $K\phi_n$ where K is any constant. As $|\phi_n|^2 \, d\tau$ represents the probability of finding the particle in the volume element $d\tau$ and as the total probability of finding the particle somewhere in space is unity, we choose the scaling constant so that

$$\int |\phi_n|^2 \, d\tau = 1 \tag{4.26}$$

Equations (4.25) and (4.26) are clearly equivalent to (4.20) so we have proved orthonormality in the case of non-degenerate eigenfunctions. We shall see later that, although degenerate eigenfunctions are not necessarily orthogonal, an orthonormal set of wave functions can always be chosen in the degenerate case also.

4.4 PROBABILITY DISTRIBUTIONS

So far we have postulated that physical quantities can be represented by Hermitian operators whose eigenvalues represent the possible results of experimental measurements and whose corresponding eigenfunctions represent the wave function of the system immediately after the measurement. Clearly if, for example, the wave function of a system is also identical with one of the energy eigenfunctions u_n immediately *before* a measurement of the energy of the system, then the result E_n will definitely be obtained; however, we have not yet postulated what the outcome of the experiment will be if the wave function before the measurement is *not* one of the u_n. The first point to be made is that, in this case, quantum mechanics does not make a precise prediction about the result of the energy measurement. This is not to say that a precise measurement of the energy is impossible (in fact Postulate 4.2 clearly states that only a result precisely equal to one of the energy eigenvalues *is* possible) but that the result of such a measurement is *unpredictable*. Clearly, the measurement of energy is just one example and similar reasoning can be applied to the measurement of other physical properties. The inability of quantum mechanics to predict the actual outcome of many individual physical events is a fundamental feature of the theory. However, the relative probabilities of the different possible outcomes can always be predicted, and we shall now proceed to develop a postulate concerning these probabilities and how they can be derived from a knowledge of the wave function and the operators representing the dynamical variables.

We have already seen (cf. Section 2.1) how a knowledge of the wave function can be used to predict the relative probabilities of the possible outcomes of a measurement of the position of a particle. We said there that if the wave function of the system before the position measurement is carried out is $\psi(\mathbf{r})$, then the probability that the position measurement will yield a result within the element of volume $d\tau$ around $\mathbf{r}$ is $|\psi(\mathbf{r})|^2 \, d\tau$. Clearly it follows that, only if the wave function has the Dirac delta form—i.e., only if it is an eigenfunction of the position operator—is the outcome certain.

We shall now generalize the above, and postulate a procedure for obtaining the probability that a result q_n will be obtained following a measurement represented by the operator $\hat{Q}$ (where $\hat{Q}\phi_n = q_n\phi_n$) on a system whose wave function is known to be ψ immediately before the measurement. The first step depends on a mathematical property of the eigenfunctions of a Hermitian operator known as *completeness*. This states that any well-behaved function, such as the wave function ψ, can be expressed as a linear combination of the eigenfunctions ϕ_n which are then said to form a *complete set*. Thus

$$\psi(\mathbf{r}) = \sum_n a_n\phi_n(\mathbf{r}) \tag{4.27}$$

where the summation is over all the eigenfunctions of the operator, and may in fact imply summing over more than one quantum number. The expression (4.27) applies to the case where the eigenvalues of the operator form a discrete set (e.g., the energy eigenvalues for a bound particle). When the eigenvalues form a continuous set (e.g., the momentum eigenvalues) indexed by the variable k instead of the integer n, the sum in (4.27) is replaced by an integral

$$\psi(\mathbf{r}) = \int a(k)\phi(k, \mathbf{r})\, dk \tag{4.28}$$

while in the general case where the set of eigenvalues consists of discrete and continuous subsets, ψ must be expressed as the sum of the right-hand sides of (4.27) and (4.28).

Two familiar examples of the application of (4.28) can be seen if we consider the expansion of an arbitrary one-dimensional wave function, $\psi(x)$, in terms of (i) the one-dimensional position eigenfunctions $\delta(x - x_0)$, and (ii) the one-dimensional momentum eigenfunctions $\exp(ikx)$. For the case (i), (4.28) becomes

$$\psi(x) = \int \psi(x_0)\delta(x - x_0)\, dx_0 \tag{4.29}$$

which clearly follows from the definitions of the Dirac delta function, where the quantity x_0 in (4.29) corresponds to k in (4.28). In case (ii) we have

$$\psi(x) = \int f(k)\, e^{ikx}\, dk$$

which is an example of the Fourier transform theorem, $f(k)$ being proportional to the Fourier transform of $\psi(x)$. Thus we have verified the correctness of the property in two particular cases. A general proof is possible, but is beyond the scope of this book.

Having accepted the mathematical statement of completeness we now have to relate it to the quantum mechanics of a physical measurement. Comparing (4.28) and (4.29) we see that if $|\psi(x)|^2\, dx$ is to represent the probability of obtaining a result between x and $x + dx$ in a position measurement, then $|a(k)|^2 \, dk$ should represent the probability of obtaining a result between $q(k)$ and $q(k + dk)$ in a measurement of a quantity whose eigenvalues form a continuous set $q(k)$. Extending this argument to the discrete case (4.27) we see that $|a_n|^2$ should represent the probability of obtaining the result q_n. We can see that this is consistent with the expected result when the wave function is identical with one of the eigenfunctions (ϕ_m, say) before the measurement and we therefore expect to obtain the result q_m with certainty; in this case it follows from (4.27) that all the a_n are zero with the exception of a_m which is unity. Thus the result q_m will certainly be obtained and the probability of obtaining any other result is zero as expected. The ideas in the previous paragraphs therefore produce the expected results in a number of special cases and we are led to state the fourth postulate.

Postulate 4.4 When a measurement of a dynamic variable represented by the Hermitian operator $\hat{Q}$ is carried out on a system whose wave function is ψ, then the probability of the result being equal to a particular discrete eigenvalue q_m will be $|a_m|^2$, where $\psi = \Sigma_n a_n \phi_n$ and the ϕ_n are the eigenfunctions corresponding to the eigenvalues q_n. In the case of continuous eigenvalues the probability of a result between $q(k)$ and $q(k + dk)$ is $|a(k)|^2 \, dk$ where $\psi(\mathbf{r}) = \int a(k)\phi(k, r) \, dk$.

We shall now proceed to use Postulate 4.4 to derive some useful general results. We shall consider explicitly the case of discrete eigenvalues only, but similar results hold in the continuous case also.

We first show that in the case where both the wave function ψ and the eigenfunctions ϕ_n are known, we can use orthonormality to obtain an expression for a_n:

$$\int \phi_n^* \psi \, d\tau = \int \phi_n^* \sum_m a_m \phi_m \, d\tau$$

$$= \sum_m a_m \int \phi_n^* \phi_m \, d\tau$$

$$= \sum_m a_m \delta_{nm} = a_n \qquad (4.30)$$

Expectation Values

Postulate 4.4 has described a procedure for calculating the relative probabilities of the results of an experiment carried out to measure a dynamical quantity on a system whose wave function is known. We can use this to predict the average value that will be obtained from a large number of measurements of the same quantity carried out on systems whose wave functions are all identical before the experiment. This average is known as the *expectation value* and is obtained in the case of a wave function ψ and an operator $\hat{Q}$ as follows. Consider the expression

$$\int \psi^* \hat{Q} \psi \, d\tau$$

Using the expansion of ψ in terms of the eigenfunctions ϕ_n of $\hat{Q}$ (4.27) we can write this as

$$\int \psi^* \hat{Q} \psi \, d\tau = \int \left(\sum_m a_m^* \phi_m^* \right) \hat{Q} \left(\sum_n a_n \phi_n \right) d\tau \qquad (4.31)$$

Thus

$$\int \psi^* \hat{Q} \psi \, d\tau = \sum_{m,n} a_m^* a_n q_n \int \phi_m^* \phi_n \, d\tau$$

$$= \sum_{m,n} a_m^* a_n q_n \delta_{mn}$$

$$= \sum_n |a_n|^2 q_n \qquad (4.32)$$

But Postulate 4.4 states that $|a_n|^2$ is just the probability that the value q_n be obtained in the measurement, so the left-hand side of (4.32) is clearly the expectation value we are looking for. This expectation value is often written as $\langle \hat{Q} \rangle$ so we have

$$\langle \hat{Q} \rangle = \int \psi^* \hat{Q} \psi \, d\tau \qquad (4.33)$$

Thus, if we know the wave function of the system and the operator representing the dynamical variable being measured, we can calculate the expectation value.

4.5 COMMUTATION RELATIONS

Consider the following expression where ψ is any wave function, and $\hat{X}$ and $\hat{P}_x$ are the operators representing the x coordinate and x component of momentum of a particle:

$$(\hat{P}_x \hat{X} - \hat{X} \hat{P}_x)\psi = -i\hbar \frac{\partial}{\partial x}(x\psi) - x\left(-i\hbar \frac{\partial \psi}{\partial x} \right)$$

$$= -i\hbar x \frac{\partial \psi}{\partial x} - i\hbar \psi + i\hbar x \frac{\partial \psi}{\partial x}$$

$$= -i\hbar \psi \qquad (4.34)$$

We note two important points. Firstly the effect on ψ of the product of two quantum mechanical operators is in general dependent on the order in which the operators are applied: that is, unlike algebraic variables, quantum mechanical operators do not in general commute. Secondly, the result (4.34) is completely independent of the particular form of ψ so we can write

$$[\hat{P}_x, \hat{X}] \equiv \hat{P}_x\hat{X} - \hat{X}\hat{P}_x = -i\hbar \qquad (4.35)$$

where $[\hat{P}_x, \hat{X}]$ is defined by (4.35) and is known as the *commutator* or *commutator bracket* of $\hat{P}_x$ and $\hat{X}$. Clearly similar arguments to the above can be used to show that

$$[\hat{P}_y, \hat{Y}] = [\hat{P}_z, \hat{Z}] = -i\hbar$$

and that commutators of the form $[\hat{X}, \hat{Y}], [\hat{P}_x, \hat{Y}], [\hat{P}_x, \hat{P}_y]$, etc., are all equal to zero. All the commutators relating to the different components of position and momentum can therefore be collected together in the following statements

$$\left.\begin{array}{l} [\hat{X}_i, \hat{X}_j] = [\hat{P}_i, \hat{P}_j] = 0 \\ [\hat{P}_i, \hat{X}_j] = -[\hat{X}_j, \hat{P}_i] = -i\hbar\,\delta_{ij} \end{array}\right\} \qquad (4.36)$$

where we are now using the notation $\hat{X}_1 \equiv \hat{X}$; $\hat{X}_2 \equiv \hat{Y}$; etc.

Expressions for the commutation relations between other pairs of operators representing dynamical variables can be obtained in a similar manner, although not every commutator is equal to a constant as was the case for $[\hat{P}_x, \hat{X}]$, etc. For example, the commutator bracket of the x coordinate of a particle and its kinetic energy $[\hat{X}, \hat{T}]$ is easily shown to be $i\hbar\hat{P}_x/m$. Commutation relations are generally of great importance in quantum mechanics. In fact it is possible to use the expressions (4.36) to define the operators representing position and momentum rather than making the explicit identifications, $\hat{X} \equiv x$ and $\hat{P}_x = -i\hbar\,\partial/\partial x$, etc. Clearly other sets of operators could be chosen which would satisfy these commutation relations; for example, the momentum operator could be put equal to the algebraic variable p with the one-dimensional position operator being $i\hbar\,\partial/\partial p$, but it can be shown that the results of quantum mechanics are independent of the choice of operators provided the commutation relations (4.36) are satisfied. It is possible to develop much of the formal theory of quantum mechanics using the commutation relations without postulating particular forms for the operators and this 'representation-free' approach is employed in many advanced texts. However, although we shall prove several important relations without using explicit expressions for the operators, we shall use the 'position representation' set up in Postulate 4.3 whenever this makes the argument clearer or more readily applicable to particular examples.

We shall now use the concept of the commutator bracket to discuss the ideas of compatible measurements and the uncertainty principle.

Compatibility

Two physical observables are said to be compatible if the operators representing them have a common set of eigenfunctions. This means that if one quantity is measured, the resulting wave function of the system will be one of the common eigenfunctions and a subsequent measurement of the other quantity will have a completely predictable result and will leave the wave function unchanged. (As will be seen later, this statement has to be modified in the degenerate case.) We shall now show that the operators representing compatible measurements commute. Let the operators be $\hat{Q}$ and $\hat{R}$ with respective eigenvalues q_n and r_n and common eigenfunctions ϕ_n. Then, if ψ is any physical wave function which can therefore be written in the form

$$\psi = \sum_n a_n \phi_n$$

we have

$$[\hat{Q}, \hat{R}]\psi = \sum_n a_n (\hat{Q}\hat{R}\phi_n - \hat{R}\hat{Q}\phi_n)$$

$$= \sum_n a_n (\hat{Q}r_n\phi_n - \hat{R}q_n\phi_n)$$

$$= \sum_n a_n (r_n q_n \phi_n - q_n r_n \phi_n)$$

$$= 0$$

Thus $[\hat{Q}, \hat{R}] = 0$ if $\hat{Q}$ and $\hat{R}$ represent compatible observables. The converse can also be proved in the non-degenerate case (the degenerate case is discussed at the end of this chapter) as we now show. Given that the operators $\hat{Q}$ and $\hat{R}$ commute, let ϕ_n be an eigenfunction of $\hat{Q}$ so that

$$\hat{Q}\hat{R}\phi_n = \hat{R}\hat{Q}\phi_n = \hat{R}q_n\phi_n = q_n\hat{R}\phi_n \tag{4.37}$$

It follows directly from (4.37) that $(\hat{R}\phi_n)$ is also an eigenfunction of $\hat{Q}$ with eigenvalue q_n. In the absence of degeneracy therefore $(\hat{R}\phi_n)$ can differ from ϕ_n only by a multiplicative constant. If we call this constant r_n we have

$$\hat{R}\phi_n = r_n\phi_n$$

and ϕ_n must therefore be an eigenfunction of both $\hat{Q}$ and $\hat{R}$. Thus, in order to test whether two quantities are compatible, it is sufficient to calculate the commutator of the operators representing them and check whether or not it is equal to zero. For example, it follows from the above and from the commutation relations (4.36) that measurements of the three positional coordinates can be made compatibly, as can measurements of the three components of momentum. Moreover, a momentum component can be measured compatibly with the measurement of either of the other two position coordinates, but not with the position coordinate in the same direction. In the case of a free particle, the energy operator ($\hat{P}^2/2m$) commutes with the momentum operator ($\hat{P}$) and these two quantities can therefore be measured compatibly: the common eigenfunctions in this case are plane waves of the form (4.7).

The Uncertainty Principle

Consider again a series of measurements of the quantity represented by the operator $\hat{Q}$ on a system whose wave function is ψ before each measurement. The average result is equal to the expectation value $\langle \hat{Q} \rangle$ which was shown earlier (4.33) to be given by

$$\langle Q \rangle = \int \psi^* \hat{Q} \psi \, d\tau$$

We can also estimate the average amount by which the result of such a measurement would be expected to deviate from this expectation value: the operator representing this 'uncertainty' is clearly ($\hat{Q} - \langle \hat{Q} \rangle$), so if the root-mean-square deviation from the mean is Δq, we have

$$\Delta q^2 = \int \psi^* (\hat{Q} - \langle \hat{Q} \rangle)^2 \psi \, d\tau$$
$$= \int \psi^* \hat{Q}'^2 \psi \, d\tau$$

where $\hat{Q}'$ is defined as ($\hat{Q} - \langle \hat{Q} \rangle$). Clearly if $\hat{Q}$ is Hermitian, so is $\hat{Q}'$ and therefore

$$\Delta q^2 = \int (\hat{Q}' \psi)(\hat{Q}'^* \psi^*)$$
$$= \int |\hat{Q}' \psi|^2 \, d\tau \tag{4.38}$$

In the same way, if we had carried out the measurement represented by the operator $\hat{R}$ on the same system, the expectation value would have been $\langle \hat{R} \rangle$ and the root-mean-square deviation Δr where

$$\Delta r^2 = \int |\hat{R}' \psi|^2 \, d\tau$$

and
$$\hat{R}' = \hat{R} - \langle \hat{R} \rangle$$

Now consider the product

$$\Delta q^2 \, \Delta r^2 = \int |\hat{Q}' \psi|^2 \, d\tau \int |R' \psi|^2 \, d\tau \; .$$
$$\geq |\int (\hat{Q}'^* \psi^*)(\hat{R}' \psi) \, d\tau|^2 \tag{4.39}$$

where the last step was obtained by an application of Schwarz's inequality which states that

$$\int |f|^2 \, d\tau \int |g^2| \, d\tau \geq |\int f^* g \, d\tau|^2$$

where f and g are any integrable functions of **r**.† Now, again using the Hermitian property of $\hat{Q}'$

$$\int(\hat{Q}'^*\psi^*)(\hat{R}'\psi)\,d\tau = \int\psi^*\hat{Q}'\hat{R}'\psi\,d\tau$$

$$= \tfrac{1}{2}\int\psi^*(\hat{Q}'\hat{R}' - \hat{R}'\hat{Q}')\psi\,d\tau$$

$$+ \tfrac{1}{2}\int\psi^*(\hat{Q}'\hat{R}' + \hat{R}'\hat{Q}')\psi\,d\tau \qquad (4.40)$$

It was shown earlier that if $\hat{Q}'$ and $\hat{R}'$ are Hermitian operators so are $(\hat{Q}'\hat{R}' + \hat{R}'\hat{Q}')$ and $i(\hat{Q}'\hat{R}' - \hat{R}'\hat{Q}')$ so the expectation values of the latter two operators are therefore real. It follows that the first term on the right-hand side of (4.40) is purely imaginary and the second term is purely real, so the right-hand side of (4.39) can be expressed as the sum of squares of the two terms on the right-hand side of (4.40) and we get

$$\Delta q^2\,\Delta r^2 \geqslant \tfrac{1}{4}(|\int\psi^*[\hat{Q}', \hat{R}']\psi\,d\tau|^2 + |\int\psi^*(\hat{Q}'\hat{R}' + \hat{R}'\hat{Q}')\psi\,d\tau|^2)$$

$$\geqslant \tfrac{1}{4}|\int\psi^*[\hat{Q}', \hat{R}']\psi\,d\tau|^2$$

Now it follows immediately from the definitions of $\hat{Q}'$ and $\hat{R}'$ that $[\hat{Q}', \hat{R}']$ $= [\hat{Q}, \hat{R}]$ so we have

$$\Delta q\,\Delta r \geqslant \tfrac{1}{2}|\int\psi^*[\hat{Q}, \hat{R}]\psi\,d\tau|$$

That is

$$\Delta q\,\Delta r \geqslant \tfrac{1}{2}|\langle[\hat{Q}, \hat{R}]\rangle| \qquad (4.41)$$

so, if we know the commutator of two operators, we can calculate the product of the uncertainties associated with series of measurements represented by each of them on a system whose wave function is ψ before each measurement. This is known as the *generalized uncertainty principle*. In the case where $\hat{Q}$ and $\hat{R}$ are the operators representing a position coordinate and a corresponding component of momentum, we have $[\hat{X}, \hat{P}_x] = i\hbar$ and therefore

$$\Delta x\,\Delta p_x \geqslant \tfrac{1}{2}\hbar \qquad (4.42)$$

which is the Heisenberg uncertainty principle referred to in Chapter 1. We

---

†To prove Schwarz's inequality, consider the expression

$$\int|f(\int|g|^2\,d\tau) - g(\int fg^*\,d\tau)|^2\,d\tau$$

This expression must be greater than or equal to zero because the integrand is nowhere negative by definition. We can rewrite the integrand as a product of a function and its complex conjugate and get:

$$\int[f^*(\int|g|^2\,d\tau) - g^*(\int f^*g\,d\tau)][f(\int|g|^2\,d\tau) - g(\int fg^*\,d\tau)]\,d\tau \geqslant 0$$

When we multiply out the product of the two square brackets, two terms cancel and we get

$$\int|f|^2\,d\tau\int|g|^2\,d\tau - |\int f^*g\,d\tau|^2 \geqslant 0$$

which is Schwarz's inequality.

note that in this particular case the uncertainty product is independent of the wave function of the system.

Thus we have shown that the uncertainty principle follows directly from the fundamental postulates of quantum mechanics. It is important to remember that the quantities Δq and Δr are not experimental errors associated with a particular measurement, but refer to the root-mean-square deviations or 'average spread' of a set of repeated measurements of physical quantities, the wave function being ψ immediately before each measurement. This point should become clearer in the following example.

Example 4.1 Single-slit diffraction Consider a beam of particles of definite momentum p travelling parallel to the z axis (i.e., the wave function is a plane wave whose wave vector has components $0, 0, p/\hbar$) towards a slit of width a in the x direction where the origin of coordinates is the centre of the slit (see Fig. 4.1). A particle has equal probability of passing through any part of the slit so the expectation value of its x coordinate at this point will be zero with a deviation Δx of about $a/2$. Beyond the slit the wave function has the form of the well-known single-slit diffraction pattern, the first minimum of which occurs at an angle θ_1 where $\sin \theta_1 = \lambda/a = 2\pi\hbar/ap$. Thus most of the particles passing through the slit travel in a direction within the central maximum of the diffraction pattern, so we can predict that measurements of the x component of the momentum of the particles will have an expectation value of zero with a spread Δp_x approximately equal to $p \sin \theta_1$. Hence

$$\Delta x \, \Delta p_x \simeq \frac{a}{2} p 2\pi \frac{\hbar}{ap} = \pi\hbar > \tfrac{1}{2}\hbar \qquad (4.43)$$

in agreement with the uncertainty principle.

We should note a number of points about the above example. First, before a particle reaches the slit it is in a momentum eigenstate in which p_x is precisely known to be zero and in which its position in the x direction is completely unknown. As the particle passes through the slit its wave function changes into one whose spread in x is Δx and whose momentum uncertainty is Δp_x. Moreover, by placing a photographic film behind the slit to record the diffraction pattern we can measure the x component of the particle momentum quite precisely by noting the point at which it blackens the film (any error in θ due to the size of the slit or to that of the blackened area can be made indefinitely small by placing the film far enough behind the slit). Hence we see that the x component of the particle momentum is precisely known before the particle enters the slit (when it is zero) and when it is recorded at the film, despite the uncertainty Δp_x calculated by the Heisenberg uncertainty principle. This is because the uncertainty Δp_x *reflects our inability*

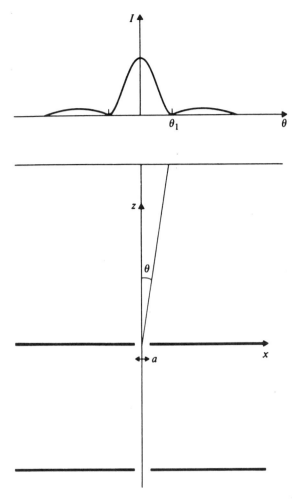

Figure 4.1 Single-slit diffraction. The lower slit ensures that particles reaching the upper slit are travelling parallel to the z axis. The graph at the top shows the intensity of the diffraction pattern as a function of θ in the case where $\lambda = a/3$.

to predict the outcome of an experimental measurement (we cannot tell in advance where on the film the particle will arrive) and *not* the accuracy of the measurement actually made.

Finally, we return to the example of the Heisenberg microscope referred to in Chapter 1 (p. 12). As a result of many photons having been scattered by the electron into the microscope we obtain an image of the electron that is 'blurred' by an amount Δx. We can therefore conclude that the electron is in a quantum state whose wave function has this width in the x direction. From the same measurement we can conclude that the uncertainty in the x

component of the electron momentum is Δp_x and we find that these two uncertainties are related by the uncertainty principle as expected. From the earlier discussion in this section, we see that what this means is that, if we were to follow up this measurement by another which determined x (say) precisely and repeat the whole process a number of times, the spread in the values of x would be Δx. Similarly, a set of momentum measurements would have a spread Δp_x.

We shall consider another example of the application of the principles of quantum mechanics to a physical system after we have discussed the fifth and final postulate.

4.6 THE TIME DEPENDENCE OF THE WAVE FUNCTION

Nearly all our discussion so far has related to the properties of wave functions immediately after measurement—i.e., to the eigenfunctions of operators representing physical quantities—and we have given very little consideration to the evolution of the wave function in time. However, at the beginning of Chapter 2 we set up the time-dependent Schrödinger equation which we generalized to three dimensions in Chapter 3 as

$$-\frac{\hbar^2}{2m}\nabla^2\Psi + V(\mathbf{r}, t)\Psi = i\hbar\frac{\partial\Psi}{\partial t} \qquad (4.44)$$

In the notation of the present chapter this becomes

$$\hat{H}\Psi = i\hbar\frac{\partial\Psi}{\partial t} \qquad (4.45)$$

where $\hat{H}$ is the Hamiltonian operator for the system. In the earlier chapters we separated off the time dependence and obtained the time-independent Schrödinger equation which we have since recognized as the energy eigenvalue equation. However, we shall now postulate that the time-dependent Schrödinger equation (4.45) must always be satisfied even when the energy of the system is not known.

> **Postulate 4.5** The development of the wave function with time is governed by the time-dependent Schrödinger equations.

We shall return to a more detailed discussion of time dependence in Chapter 8, but in the meantime we discuss only the particular case where $\hat{H}$ is not itself a function of time; classically this corresponds to a closed system in which energy is conserved. If the energy eigenfunctions of the system obtained by solving the energy eigenvalue equation (4.2) are u_n then, by completeness, the wave function at any time t can be expressed as a linear combination of the u_n

$$\psi(\mathbf{r}, t) = \sum_n a_n(t) u_n(\mathbf{r}) \tag{4.46}$$

where the coefficients a_n are in general functions of time. Substituting (4.46) into (4.45) we get

$$i\hbar \sum_n \frac{da_n}{dt} u_n = \sum_n a_n \hat{H} u_n$$

$$= \sum_n a_n E_n u_n$$

Thus

$$\sum_n \left(i\hbar \frac{da_n}{dt} - a_n E_n \right) u_n = 0 \tag{4.47}$$

Equation (4.47) must be true at all points in space so the terms in brackets must vanish leading to

$$i\hbar \frac{da_n}{dt} = a_n E_n$$

that is,

$$a_n(t) = a_n(0) \exp(-iE_n t/\hbar) \tag{4.48}$$

where $a_n(0)$ is the value of a_n at some initial time $t = 0$. Hence from (4.46) and (4.48) a general solution to the time-dependent Schrödinger equation is

$$\Psi(\mathbf{r}, t) = \sum_n a_n(0) u_n(\mathbf{r}) \exp(-iE_n t/\hbar) \tag{4.49}$$

Now suppose we carry out a measurement of the energy of the system at time $t = 0$. From the earlier postulates we should obtain a result equal to one of the energy eigenvalues, say E_m. Moreover, the wave function immediately after the measurement will be the corresponding eigenfunction u_m. This is equivalent to saying that at $t = 0$, $a_m(0) = 1$ and $a_n(0) = 0$, $n \neq m$. In this case (4.49) becomes

$$\Psi(\mathbf{r}, t) = u_m \exp(-iE_m t/\hbar) \tag{4.50}$$

which was the particular form of solution used in Chapters 2 and 3. We notice that the right-hand side of (4.50) differs from u_m only by a phase factor and so Ψ is an eigenfunction of $\hat{H}$ at all times (remember $\hat{H}$ is assumed to be time independent). Thus any later measurement of the energy will again yield the value E_m and we conclude that energy is conserved in such a quantum system just as it would be classically. Moreover, the value of any quantity that can be measured compatibly with the energy and whose operator therefore commutes with $\hat{H}$ (e.g., the linear momentum of a free particle) will also be conserved. The above result emphasizes the great importance of the time-independent Schrödinger equation (the energy eigenvalue equation) whose solutions we discussed in some detail in earlier chapters. Much of quantum mechanics concerns the properties of the energy eigenstates of systems whose Hamiltonians are independent of time—often referred to as 'stationary states' because of the properties described above. On the other hand many experimental measurements (e.g. atomic spectra) refer to systems which change from one nearly stationary state to another under the influence of a time-dependent potential. We shall discuss the quantum mechanics of such changes in Chapter 9 where we shall see that in many cases the time-dependent potential can be considered as causing small 'perturbations' on the stationary states of the system.

4.7 THE MEASUREMENT OF MOMENTUM BY COMPTON SCATTERING

An example of a real physical measurement that illustrates many of the principles discussed in this chapter is the measurement of the momentum of electrons by Compton scattering. We discussed Compton scattering in Chapter 1 where we obtained evidence for the existence of photons of energy $\hbar\omega$ and momentum $\hbar\mathbf{k}$ where ω and $\mathbf{k}$ are the angular frequency and wave vector of the X-rays. However, at that stage we assumed that the electron was at rest before the scattering event; we will now lift this restriction and show how Compton scattering can be used to obtain information about the momentum of the scattering electron.

We consider first the case of a free electron whose momentum before and after the photon is scattered has the values $\mathbf{p}$ and $\mathbf{p}'$ respectively and treat this situation, as in Chapter 1, using the classical laws governing the collisions of particles. Measurement of the wavelength of the incident and scattered X-rays (e.g. by Bragg reflection from a crystal of known lattice spacing) and knowledge of the direction of the incoming and outgoing photon lead to values for the change in momentum ($\Delta\mathbf{P}$) and the change in energy (ΔE) of the photon in the collision. Using conservation of energy and momentum we have

$$\Delta\mathbf{P} = \mathbf{p} - \mathbf{p}' \qquad (4.51)$$

and

$$\Delta E = p^2/2m - p'^2/2m \qquad (4.52)$$

Squaring (4.51) we get

$$p'^2 = p^2 + \Delta P^2 - 2\mathbf{p}\cdot\Delta\mathbf{P} \qquad (4.53)$$

while from (4.52) we have

$$p'^2 = p^2 - 2m\,\Delta E \qquad (4.54)$$

Equating the right-hand sides of (4.53) and (4.54) leads to

$$2\mathbf{p}\cdot\Delta\mathbf{P} = \Delta P^2 + 2m\,\Delta E \qquad (4.55)$$

Thus our knowledge of $\Delta\mathbf{P}$ and ΔE have led directly to a measurement of the component of $\mathbf{p}$ in the direction of $\Delta\mathbf{P}$ and a similar component of $\mathbf{p}'$ is readily derived from (4.55) and (4.51). We now analyse this measurement from the point of view of quantum mechanics and note again that for a free particle $(V = 0)$ the momentum operator $(-i\hbar\nabla)$ commutes with the Hamiltonian $(-\hbar^2\nabla^2/2m)$ so these are compatible measurements and the momentum eigenfunctions are therefore stationary states of the system. The experiment then implies that the electron was in a state of momentum $\mathbf{p}$ (wave function $\exp(i\mathbf{p}\cdot\mathbf{r}/\hbar)$) before the measurement and in another stationary state of momentum $\mathbf{p}'$ (wave function $\exp(i\mathbf{p}'\cdot\mathbf{r}/\hbar)$) afterwards. The outcome of the experiment is therefore completely predictable and if it is repeated with the same starting conditions the same result will certainly be obtained for the component of $\mathbf{p}'$ in the direction $\Delta\mathbf{P}$.

The situation becomes more complicated, but at the same time more illustrative of the quantum theory of measurement, if the electron is bound to an atom before the scattering process takes place. We shall assume that a previous measurement of the energy of the atom (achieved perhaps by observing photons emitted from it) has shown it to be in its ground state which is of course a stationary state of the system. We now consider the dynamics of the scattering process. The conservation of energy and momentum is complicated by the presence of the atomic nucleus, and, unless the atom is ionized in the collision, the momentum transfer will be to the atom as a whole and the experiment will give little or no information about the momentum of the individual electron. Even if the atom is ionized, Eq. (4.55) will provide an accurate measurement of the component of the electron momentum only if the energy transfer is much larger than the ionization energy: that is if the electron is knocked 'cleanly' out of the atom. Thus in order to measure accurately the momentum of an electron in an atom, we must make the electron effectively free in the process. Looking at this experiment from the point of view of quantum mechanics we first note that, if the electron is bound to the atom, the momentum operator does not commute with the Hamiltonian operator so the initial state cannot be a momentum eigenstate. As we know the change in photon energy, we can use conservation of energy to see that the atom must be in an energy eigenstate after the collision, but if the momentum has been measured it follows from Postulate 4.2 that the electron is also in a momentum eigenstate after the measurement. The only way both these conditions can be fulfilled is if the electron is effectively free after the collision which is just the condition we derived above as necessary for an accurate momentum measurement.

We have seen, therefore, that an accurate measurement of the momentum of an electron in an atom is possible provided the electron is placed in a momentum eigenstate as a result of the experiment. What is not possible on the basis of the postulates of quantum mechanics is a prediction of the precise result of such an experiment. Postulate 4.4, however, tells us how to obtain the relative probabilities of the outcome of such experiments and we complete the discussion by applying this to the present example. We assume the atom is hydrogen and that it is initially in its ground state. Then its wave function is, from Chapter 3,

$$u(\mathbf{r}) = (\pi a_0^3)^{1/2} \exp(-r/a_0) \tag{4.56}$$

The momentum eigenvalues form a continuous set so the probability of obtaining a value of the momentum in the region $dp_x \, dp_y \, dp_z$ around $\mathbf{p}$ is given by

$$|f(\mathbf{k})|^2 \, dk_x \, dk_y \, dk_z$$

where

$$u(\mathbf{r}) = \iiint f(\mathbf{k}) \exp(i\mathbf{k} \cdot \mathbf{r}) \, dk_x \, dk_y \, dk_z \tag{4.57}$$

and

$$k_x = p_x/\hbar \qquad \text{etc.}$$

remembering that $\exp(i\mathbf{k} \cdot \mathbf{r})$ is the momentum eigenfunction corresponding to the eigenvalue $\hbar\mathbf{k}$. Thus $f(\mathbf{k})$ is just the Fourier transform of $u(\mathbf{r})$ and we can perform the reverse transform leading to

$$f(\mathbf{k}) = \left(\frac{1}{8\pi^3}\right)^{1/2} \int u(\mathbf{r}) \exp(-i\mathbf{k} \cdot \mathbf{r}) \, d\tau$$

$$= (8\pi^4 a_0^3)^{-1/2} \int_0^{2\pi} \int_0^{\pi} \int_0^{\infty} \exp(-r/a_0 - ikr \cos\theta) r^2 \, dr \sin\theta \, d\theta \, d\phi \tag{4.58}$$

where we have substituted from (4.56) into (4.57) and expressed the volume integral in spherical polar coordinates with the direction of $\mathbf{k}$ as the polar axis. The integrals over θ and ϕ are readily evaluated after making the substitution $x = \cos\theta$ leading to

$$f(\mathbf{k}) = 2(2\pi^2 a_0^3)^{-1/2} \int_0^{\infty} k^{-1} \sin kr \exp(-r/a_0) r \, dr$$

The integral over r can be evaluated by successive integrations by parts and we obtain

$$f(\mathbf{k}) = 2(2a_0^3)^{1/2}/[\pi(1 + a_0^2 k^2)^2] \tag{4.59}$$

As expected, therefore, the momentum probability distribution for an electron in the ground state of the hydrogen atom is spherically symmetric in momentum space, corresponding with the spherically symmetric position probability distribution obtained by squaring (4.56). Finally we note that the Compton scattering experiment measures one component of momentum only. Calling this the z component, the probability that it will be found to have a value between $\hbar k_z$ and $\hbar(k_z + dk_z)$ while the other components may adopt any values will be $P(k_z)\, dk_z$ where

$$P(k_z) = \int \int |f(\mathbf{k})|^2 \, dk_x \, dk_y$$

$$= (8a_0^3/\pi^2) \int_{-\infty}^{\infty} \int_{-\infty}^{\infty} (1 + a_0^2 k_x^2 + a_0^2 k_y^2 + a_0^2 k_z^2)^{-4} \, dk_x \, dk_y$$

The relevant integrals can be looked up in tables giving

$$P(k_z) = 2a_0/[\pi(1 + a_0^2 k_z^2)^2] \tag{4.60}$$

We note that the product of the widths of the momentum distribution $(\Delta p_z \simeq \hbar/a_0)$ and the position distribution $(\Delta z \simeq a_0)$ is approximately equal to $\hbar$ in agreement with the uncertainty principle.

We have discussed the above example in considerable detail because it illustrates several of the main features of the basic postulates of quantum mechanics. In particular we have seen how an exact measurement of a physical quantity (in this case the momentum of an electron) results in the system being in an eigenstate of the operator representing the measurement immediately after the measurement has been completed. We have also seen that a knowledge of the wave function of the system before the measurement does not generally provide a precise prediction of the result, but only of the relative probabilities of the various outcomes.

Another reason for discussing this particular example is that Compton scattering represents a measurement which can be actually carried out. Unfortunately, such experiments on monatomic hydrogen are impractical, but Fig. 4.2 shows the momentum component probability distribution in helium, calculated in a similar manner to the above, and also the results of Compton scattering experiments, each experimental point representing the number of times the component of momentum of an electron was measured to be in the region of that value. We see that there is excellent agreement between the frequency distribution and the calculated probability distribution, but we note once more that quantum mechanics can only make this statistical prediction and that the values obtained in individual measurements of electron momentum are unpredictable.

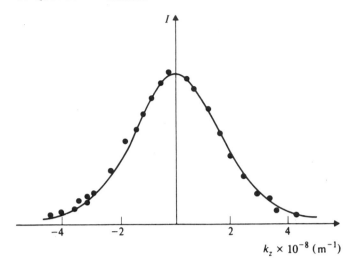

Figure 4.2 A comparison of the theoretical and experimental intensities of Compton scattering from helium. The continuous line represents the momentum probability distribution calculated in the manner described in the text and the points represent measured intensities—after J. W. M. Dumond and H. A. Kirkpatrick, *Physical Review*, vol. 52, pp. 419–436, 1937.

4.8 DEGENERACY

The previous sections of this chapter apply to cases where the eigenfunctions of the operators representing physical measurements are non-degenerate. That is, we have assumed that if

$$\hat{Q}\phi_n = q_n\phi_n$$

then $\qquad\qquad q_n \neq q_m \qquad$ for all $n \neq m$ $\qquad\qquad$ (4.61)

However, degeneracy is in fact quite a common feature of physical systems, and indeed we saw in Chapter 3 that it was a necessary consequence of the symmetry of many three-dimensional systems. In the present section, therefore, we shall extend our discussion of the formal properties of quantum mechanics to degenerate systems. There have been two occasions in which we have explicitly assumed that systems were non-degenerate: first when discussing orthonormality and second when discussing compatible measurements. We shall now extend our discussion of both these points to include degeneracy.

We first show that any linear combination of degenerate eigenfunctions (ϕ_n) with the same eigenvalue (q) is also an eigenfunction with that eigenvalue. This follows directly from substitution into (4.61)

$$\hat{Q} \sum_n c_n \phi_n = \sum_n c_n \hat{Q} \phi_n$$

$$= q \sum_n c_n \phi_n \qquad (4.62)$$

A complete set of degenerate eigenfunctions is the minimum number such that any other eigenfunction with this eigenvalue can be expanded as a linear combination of the members of this set. We shall now show that, although degenerate eigenfunctions need not be orthogonal, it is always possible to construct a complete set of orthogonal eigenfunctions, given a complete set of non-orthogonal eigenfunctions, ϕ_n.

Consider the function ϕ_2' where

$$\phi_2' = S_{12}\phi_1 - \phi_2 \qquad \text{and} \qquad S_{12} = \int \phi_1^* \phi_2 \, d\tau$$

then

$$\int \phi_1^* \phi_2' \, d\tau - S_{12} \int \phi_1^* \phi_1 \, d\tau - \int \phi_1^* \phi_2 \, dt$$

$$= S_{12} - S_{12} = 0$$

Thus ϕ_2' is orthogonal to ϕ_1 and a third function orthogonal to these two can be similarly shown to be ϕ_3' where

$$\phi_3' = S_{13}\phi_1 + S_{23}\phi_2' - \phi_3$$

and

$$S_{23} = \int \phi_2'^* \phi_3 \, d\tau / \int |\phi_2'|^2 \, d\tau$$

This procedure (known as *Schmidt orthogonalization*) can be continued until a complete orthogonal set of degenerate eigenfunctions has been set up. Thus all the results in this chapter which depend on orthogonality can be extended to the degenerate case, provided it is assumed that such an orthogonal set has first been constructed.

Turning now to a discussion of compatible measurements, the proof of the theorem that the operators representing compatible measurements commute is clearly unaffected by the presence of degeneracy, but the converse (that measurements represented by commuting operators are compatible) requires modification. Consider the case where ϕ_1 and ϕ_2 are two degenerate eigenfunctions (eigenvalue q) of the operator $\hat{Q}$ which commutes with the operator $\hat{R}$. Assume for the moment that there are only two linearly independent eigenfunctions with this eigenvalue. Then

$$\hat{Q}\hat{R}\phi_1 = \hat{R}\hat{Q}\phi_1 = q\hat{R}\phi_1 \qquad (4.63)$$

so $(\hat{R}\phi_1)$ is also an eigenfunction of $\hat{Q}$ with eigenvalue q, and it must therefore be a linear combination of ϕ_1 and ϕ_2. That is,

$$\hat{R}\phi_1 = a\phi_1 + b\phi_2 \qquad (4.64)$$

where a and b are constants. An identical argument, replacing ϕ_1 with ϕ_2 in (4.63), leads to

$$\hat{R}\phi_2 = c\phi_1 + d\phi_2 \tag{4.65}$$

where c and d are constants. We shall now show that particular linear combinations of ϕ_1 and ϕ_2 exist that are eigenfunctions of $\hat{R}$. Let one of these be $\phi' = A\phi_1 + B\phi_2$ and let its eigenvalue be r, then

$$\hat{R}\phi' \equiv A\hat{R}\phi_1 + B\hat{R}\phi_2 \equiv Aa\phi_1 + Ab\phi_2 + Bc\phi_1 + Bd\phi_2$$
$$= r\phi' \equiv rA\phi_1 + rB\phi_2 \tag{4.66}$$

Equating coefficients of ϕ_1 and ϕ_2 in (4.66) we get

$$\left.\begin{array}{r} (a - r)A + cB = 0 \\ bA + (d - r)B = 0 \end{array}\right\} \tag{4.67}$$

The equations (4.67) have solutions only if the determinant of the coefficients of A and B vanishes, leading to two possible values of r. In general these will not be equal, so the eigenfunctions of $\hat{R}$ need not be degenerate. This result can readily be extended to the case of an arbitrary number of degenerate functions. We therefore conclude that, although all possible forms of the degenerate eigenfunctions of one operator are not necessarily eigenfunctions of the other, a set of eigenvectors that are common to both operators can always be chosen.

Finally we consider the implications of the above for the quantum theory of measurement. Following a measurement represented by the operator $\hat{Q}$, which produces the result q, the system will have a wave function equivalent to one of the set of degenerate eigenfunctions that correspond with this eigenvalue. However, this need not be an eigenfunction of the operator $\hat{R}$ and the exact result of a measurement represented by this operator is not predictable even though $\hat{Q}$ and $\hat{R}$ commute. Nevertheless, the only possible results are those calculable in the manner described above, and a subsequent measurement of the quantity represented by $\hat{R}$ will leave q unchanged and result in the wave function being equivalent to one of the set of common eigenfunctions so that the results of further measurements represented by $\hat{Q}$ or $\hat{R}$ will be completely predictable.

As an example of this, consider the energy eigenfunctions of the hydrogen atom discussed in Chapter 3. A measurement of the energy that showed the atom to have principal quantum number $n = 2$ would result in the atom being in one of four degenerate states ($l = 0, m = 0; l = 1, m = 0$ or ± 1) or in some linear combination of them. A subsequent measurement of the total angular momentum must yield a result $\sqrt{l(l + 1)}\hbar$ where l is either zero or one, but which of these is unpredictable. Following this, if l has been found to be equal to one, say, the value of m will be similarly unpredictable unless the z component of angular momentum is also measured. However, once all three measurements have been made, the wave function of the system is completely specified and the results of further measurements of any of these quantities are completely predictable.

PROBLEMS

4.1 Show that the operators $\hat{Q}, \hat{Q}^2, \hat{Q}^3$, etc., are all compatible. Hence show that any component of the momentum of a particle can always be measured compatibly with the kinetic energy, but that the momentum and total energy can be measured compatibly only if the potential energy is constant everywhere.

4.2 A particle moves in one dimension subject to a potential that is zero in the region $-a \leqslant x \leqslant a$ and infinite elsewhere. At a certain time its wave function is

$$\psi = (5a)^{-1/2} \cos(\pi x/2a) + 2(5a)^{-1/2} \sin(\pi x/a)$$

What are the possible results of the measurement of the energy of this system and what are their relative probabilities? What is the form of the wave function immediately after such a measurement? If the energy is immediately remeasured, what will now be the relative probabilities of the possible outcomes? (The energy eigenvalues and eigenfunctions of this system are given in Sec. 2.4.)

4.3 The energy of the particle in Prob. 4.2 is measured and a result equal to the lowest energy eigenvalue is obtained. Show that the probability of a subsequent measurement of the electron momentum yielding a result between $\hbar k$ and $\hbar(k + dk)$ is equal to $P(k) dk$ where

$$P(k) = \frac{\pi}{2a^3} \frac{\cos^2 ka}{(\pi^2/4a^2 - k^2)^2}$$

4.4 Show that if the particle in Prob. 4.2 had been in a highly excited energy eigenstate of eigenvalue E, a measurement of the momentum would almost certainly have produced a value equal to $\pm(2mE)^{1/2}$. Compare this result with the predictions of classical mechanics for this problem.

4.5 A particle is observed to be in the lowest energy state of an infinite-sided well. The width of the well is (somehow) expanded to double its size so quickly that the wave function does not change *during this process*; the expansion takes place symmetrically so that the centre of the well does not move. The energy of the particle is measured again. Calculate the probabilities that it will be found in (i) the ground state, (ii) the first excited state, and (iii) the second excited state of the expanded well.

4.6 Calculate the expectation values of (i) x, (ii) x^2, (iii) p, and (iv) p^2 for a particle known to be in the lowest energy state of a one dimensional harmonic oscillator potential before these measurements are performed. Hence show that the expectation value of the total energy of the particle equals the ground-state eigenvalue and that the product of the root-mean-square deviations of x and p from their respective means has the minimum value allowed by the uncertainty principle.

4.7 Calculate the expectation values of the following quantities for an electron known to be in the ground state of the hydrogen atom before the measurements are performed: (i) the distance r of the electron from the nucleus, (ii) r^2, (iii) the potential energy, (iv) the kinetic energy. Show that the sum of (iii) and (iv) equals the total energy.

4.8 An operator $\hat{P}$, known as the parity operator, is defined so that, in one dimension, $\hat{P}f(x) = f(-x)$ where f is any well-behaved function of x. Assuming that $\hat{P}$ is real, prove that $\hat{P}$ is Hermitian. Show that the eigenvalues of $\hat{P}$ are ± 1 and that any function $\phi(x)$ that has definite parity—i.e., if $\phi(x) = \pm\phi(-x)$—is an eigenfunction of $\hat{P}$.

4.9 Show that, if H is the Hamiltonian operator representing the total energy of a system where the potential energy is centrosymmetric and if $\hat{P}$ is the parity operator defined in Prob. 4.8, then $[\hat{P}, \hat{H}] = 0$. Hence show that in the non-degenerate case, the energy eigenfunctions of such a system always have definite parity. What restrictions, if any, are placed on the eigenfunctions in the similar, but degenerate, case?

4.10 Generalize the results given in Probs 4.8 and 4.9 to the three-dimensional case and confirm that they apply to the energy eigenfunctions of a particle in a spherically symmetric potential obtained in Chapter 3.

4.11 X-rays of wavelength 1.00×10^{-10} m are incident on a target containing free electrons, and a Compton scattered X-ray photon of wavelength 1.02×10^{-10} m is detected at an angle of 90° to the incident direction. Obtain as much information as possible about the momentum of the scattering electron before and after the scattering process.

FIVE

ANGULAR MOMENTUM I

Angular momentum is a very important and revealing property of many physical systems. In classical mechanics the principle of conservation of angular momentum is a powerful aid to the solution of such problems as the motion of planets and satellites and the behaviour of gyroscopes and tops. The role of angular momentum in quantum mechanics is probably even more important and this will be the subject of the present chapter and the next. We shall find that the operators representing the components of angular momentum do not commute with each other, although they all commute with the operator representing the total angular momentum. It follows that no pair of these components can be measured compatibly and we shall therefore look for a set of eigenfunctions that are common to the operators representing the total angular momentum and one of its components. We shall find that the angular momentum eigenvalues always form a discrete set and that, in the case of a central field, the operators commute with the Hamiltonian, implying that the total angular momentum and one component can be measured compatibly with the energy in this case, and that their values, once measured, remain constant in time. Consideration of angular momentum will also enable us to make predictions about the behaviour of atoms in magnetic fields, and we shall find that these are in agreement with experiment only if we assume that the electron (along with other fundamental particles) has an intrinsic 'spin' angular momentum in addition to the 'orbital' angular momentum associated with its motion; the interaction between spin and orbital angular momenta leads to detailed features of the atomic spectra which are known as 'fine structure'. We shall also use the measurement of angular momentum to illustrate the quantum theory of measurement discussed in the previous chapter.

Angular momentum is therefore an important subject which deserves and requires detailed consideration. In the rest of this chapter we shall consider the properties of the angular-momentum operators and obtain expressions for their eigenvalues in the case of spin as well as orbital angular momentum, while in the next chapter we shall show that the angular-momentum components can be usefully represented by matrices rather than differential operators, and discuss how this matrix representation can be used to describe the fine structure of atomic spectra and to illustrate the quantum theory of measurement.

5.1 THE ANGULAR-MOMENTUM OPERATORS

The classical expression for the angular momentum $\mathbf{l}$ of a particle whose position and momentum are $\mathbf{r}$ and $\mathbf{p}$ respectively is

$$\mathbf{l} = \mathbf{r} \times \mathbf{p} \tag{5.1}$$

The third postulate discussed in the previous chapter states that operators representing dynamical quantities bear the same functional relationship to the position and momentum operators as do the corresponding classical quantities to the classical position and momentum variables. It follows therefore that the quantum-mechanical operator, $\hat{\mathbf{L}}$, representing angular momentum is

$$\hat{\mathbf{L}} = \hat{\mathbf{R}} \times \hat{\mathbf{P}} \tag{5.2}$$

where $\hat{\mathbf{R}}$ and $\hat{\mathbf{P}}$ are the operators representing position and momentum. The operators representing the Cartesian components of angular momentum are therefore

$$\hat{L}_x = \hat{Y}\hat{P}_z - \hat{Z}\hat{P}_y \qquad \hat{L}_y = \hat{Z}\hat{P}_x - \hat{X}\hat{P}_z \qquad \hat{L}_z = \hat{X}\hat{P}_y - \hat{Y}\hat{P}_x \tag{5.3}$$

and that representing the square of the magnitude of the total angular momentum is

$$\hat{L}^2 = \hat{L}_x^2 + \hat{L}_y^2 + \hat{L}_z^2 \tag{5.4}$$

Before discussing the eigenvalues and eigenfunctions of these operators, we should find out whether they represent quantities that are compatible in the sense discussed in the previous chapter. This implies that we should discover whether the operators $\hat{L}_x$, $\hat{L}_y$, $\hat{L}_z$, and $\hat{L}^2$ commute when taken in pairs. We do this using (5.3), (5.4), and the commutation relations (4.36) for the position and momentum components:

$$[\hat{L}_x, \hat{L}_y] = \hat{L}_x\hat{L}_y - \hat{L}_y\hat{L}_x$$
$$= (\hat{Y}\hat{P}_z - \hat{Z}\hat{P}_y)(\hat{Z}\hat{P}_x - \hat{X}\hat{P}_z) - (\hat{Z}\hat{P}_x - \hat{X}\hat{P}_z)(\hat{Y}\hat{P}_z - \hat{Z}\hat{P}_y)$$
$$= \hat{Y}\hat{P}_z\hat{Z}\hat{P}_x - \hat{Y}\hat{P}_z\hat{X}\hat{P}_z - \hat{Z}\hat{P}_y\hat{Z}\hat{P}_x + \hat{Z}\hat{P}_y\hat{X}\hat{P}_z - \hat{Z}\hat{P}_x\hat{Y}\hat{P}_z + \hat{Z}\hat{P}_x\hat{Z}\hat{P}_y$$
$$+ \hat{X}\hat{P}_z\hat{Y}\hat{P}_z - \hat{X}\hat{P}_z\hat{Z}\hat{P}_y$$

Remembering the commutation relations for the components of position and momentum (4.36) we see that the second and third terms in the expression for $[\hat{L}_x, \hat{L}_y]$ cancel with the sixth and seventh terms while the others can be rearranged to give

$$[\hat{L}_x, \hat{L}_y] = (\hat{Y}\hat{P}_x - \hat{X}\hat{P}_y)(\hat{P}_z\hat{Z} - \hat{Z}\hat{P}_z)$$
$$= i\hbar(\hat{X}\hat{P}_y - \hat{Y}\hat{P}_x)$$
$$= i\hbar\hat{L}_z$$

$[\hat{L}_y, \hat{L}_z]$ and $[\hat{L}_z, \hat{L}_x]$ can be obtained similarly and we have

$$[\hat{L}_x, \hat{L}_y] = i\hbar\hat{L}_z \qquad (5.5)$$

$$[\hat{L}_y, \hat{L}_z] = i\hbar\hat{L}_x \qquad (5.6)$$

$$[\hat{L}_z, \hat{L}_x] = i\hbar\hat{L}_y \qquad (5.7)$$

We can now complete our calculation of the commutator brackets by considering those involving $\hat{L}^2$. From (5.4)

$$[\hat{L}^2, \hat{L}_z] = [\hat{L}_x^2, \hat{L}_z] + [\hat{L}_y^2, \hat{L}_z] + [\hat{L}_z^2, \hat{L}_z] \qquad (5.8)$$

Considering the first term on the right-hand side and using equations (5.5) to (5.7)

$$[\hat{L}_x^2, \hat{L}_z] = \hat{L}_x\hat{L}_x\hat{L}_z - \hat{L}_z\hat{L}_x\hat{L}_x$$
$$= \hat{L}_x(-i\hbar\hat{L}_y + \hat{L}_z\hat{L}_x) - (i\hbar\hat{L}_y + \hat{L}_x\hat{L}_z)\hat{L}_x$$
$$= -i\hbar(\hat{L}_x\hat{L}_y + \hat{L}_y\hat{L}_x) \qquad (5.9)$$

Similarly

$$[\hat{L}_y^2, \hat{L}_z] = i\hbar(\hat{L}_x\hat{L}_y + \hat{L}_y\hat{L}_x) \qquad (5.10)$$

Also

$$[\hat{L}_z^2, \hat{L}_z] = \hat{L}_z^3 - \hat{L}_z^3 = 0 \qquad (5.11)$$

Substituting from (5.9), (5.10), and (5.11) into (5.8) and generalizing the result to the commutator brackets containing the other components we get

$$[\hat{L}^2, \hat{L}_x] = [\hat{L}^2, \hat{L}_y] = [\hat{L}^2, \hat{L}_z] = 0 \qquad (5.12)$$

It follows from (5.5) to (5.7) that the operators representing any two components of angular momentum do not commute and are therefore not compatible. If, therefore, the system is in an eigenstate of one angular momentum component, it will not be simultaneously in an eigenstate of either of the others. On the other hand, (5.12) shows that the total angular momentum *can* be measured compatibly with any one component. We shall therefore approach the eigenvalue problem by looking for a set of functions which are simultaneously eigenfunctions of the total angular momentum and *one* component, which is conventionally chosen to be the z component $\hat{L}_z$.

5.2 THE EIGENVALUES AND EIGENFUNCTIONS

We can obtain an explicit expression for the operators representing the angular momentum components using the forms of the position and momentum operators given in Postulate 4.3:

$$\hat{P} = -i\hbar\nabla \qquad \hat{R} = r$$

Hence

$$\hat{L} = -i\hbar r \times \nabla \tag{5.13}$$

When considering the angular momentum of a particle moving about a point, it is a great advantage to refer to a spherical polar coordinate system similar to that used in Chapter 3 when discussing the energy eigenvalues of a particle in a central field. Expressions for $\hat{L}_z$ and $\hat{L}^2$ in this coordinate system will be derived using vector calculus, but readers who are not familiar with this technique may prefer to accept Eqs (5.16) and (5.17) below and proceed to the discussion following these equations.

We assume that the vector **r** is referred to the origin of the spherical polar coordinate system, giving us the standard expression

$$\nabla = r_0 \frac{\partial}{\partial r} + \frac{1}{r}\theta\frac{\partial}{\partial\theta} + \frac{1}{r\sin\theta}\phi\frac{\partial}{\partial\phi} \tag{5.14}$$

where r_0, θ, and ϕ are the three basic unit vectors of the spherical polar system. It follows directly that

$$\hat{L} = -i\hbar r \times \nabla = -i\hbar\left(\phi\frac{\partial}{\partial\theta} - \frac{1}{\sin\theta}\theta\frac{\partial}{\partial\phi}\right) \tag{5.15}$$

Taking the polar axis to be z, the unit vector, z_0, in this direction is

$$z_0 = \cos\theta r_0 - \sin\theta\theta$$

Hence

$$\hat{L}_z = \mathbf{z}_0 \cdot \hat{\mathbf{L}} = -i\hbar \frac{\partial}{\partial \phi} \tag{5.16}$$

An expression for $\hat{L}^2$ is obtained from

$$\hat{L}^2 = -\hbar^2 (\mathbf{r} \times \nabla) \cdot (\mathbf{r} \times \nabla)$$
$$= -\hbar^2 \mathbf{r} \cdot \nabla \times (\mathbf{r} \times \nabla)$$

Substituting for $\mathbf{r} \times \nabla$ from (5.15) this expression becomes

$$\hat{L}^2 = -\hbar^2 \mathbf{r} \cdot \nabla \times \left(\boldsymbol{\phi} \frac{\partial}{\partial \theta} - \boldsymbol{\theta} \frac{1}{\sin \theta} \frac{\partial}{\partial \phi} \right)$$

so that, using the standard expression for curl in spherical polar coordinates we get

$$\hat{L}^2 = \hbar^2 \left[\frac{1}{\sin \theta} \frac{\partial}{\partial \theta} \left(\sin \theta \frac{\partial}{\partial \theta} \right) + \frac{1}{\sin^2 \theta} \frac{\partial^2}{\partial \phi^2} \right] \tag{5.17}$$

We can now use (5.16) and (5.17) to obtain the eigenvalues and the common set of eigenfunctions for the operators $\hat{L}_z$ and $\hat{L}^2$. Considering $\hat{L}^2$ first, we notice that (5.17) is identical (apart from a multiplicative constant) to the differential equation (3.26) determining the angularly dependent part of the energy eigenfunctions in the central field problem which has already been solved in Chapter 3. It follows directly from that solution (cf. (3.44)) that the eigenvalues of $\hat{L}^2$ are

$$l(l+1)\hbar^2 \tag{5.18}$$

where l is a positive integer or zero, and that the corresponding eigenfunctions are

$$Y_{lm}(\theta, \phi) = (-1)^m \left[\frac{(2l+1)}{4\pi} \frac{(l-|m|)!}{(l+|m|)!} \right]^{1/2} P_l^{|m|}(\cos \theta) \, e^{im\phi} \tag{5.19}$$

where m is an integer whose modulus is less than or equal to l, and some of the properties of the spherical harmonics Y_{lm} and the associated Legendre functions, $P_l^{|m|}$, have been discussed in Chapter 3. It follows immediately from (5.16) and (5.19) that

$$\hat{L}_z Y_{lm} = -i\hbar \frac{\partial Y_{lm}}{\partial \phi} = m\hbar Y_{lm} \tag{5.20}$$

so the functions Y_{lm} are the simultaneous eigenfunctions of $\hat{L}_z$ and $\hat{L}^2$ that we have been looking for and the eigenvalues of $\hat{L}_z$ are given by

$$m\hbar \qquad \text{where} \quad -l \leqslant m \leqslant l \tag{5.21}$$

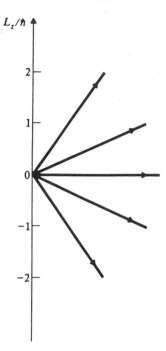

Figure 5.1 The possible orientations of the angular momentum vector relative to the z axis in the case where $l = 2$. Note that the orientation in the xy plane is not known if the z component has a definite value.

We have therefore confirmed the physical interpretation of the quantum numbers l and m that we anticipated in Chapter 3: the square of the total angular momentum is quantized in units of $l(l + 1)\hbar^2$ and the z component in units of $m\hbar$. The condition $-l \leqslant m \leqslant l$ follows directly from the obvious requirement that a measurement of the total angular momentum must yield a result at least as large as that from a simultaneous measurement of any component. If the magnitude of the total angular momentum and that of the z component are known, it follows that the angular momentum vector must be at a fixed angle to the z axis. However, its orientation with respect to the x and y axes is unknown and neither of these components can therefore be measured compatibly with $\hat{L}_z$, which is consistent with the results derived above from the commutation relations. The possible orientations with respect to the z axis are illustrated in Fig. 5.1 for the case where $l = 2$: we note that there is no eigenstate where the angular momentum is exactly parallel to the z axis, which is to be expected as this configuration would imply that the x and y components were then both known to be exactly equal to zero.

Our knowledge of the angular-momentum eigenvalues and eigenfunctions also helps us to understand more clearly the physical basis of the degeneracy found in the energy states of the central-field problem in Chapter 3. The central-field Hamiltonian, which is contained in the time-independent Schrödinger equation (3.23), can be written using (5.17) in the form

$$\hat{H} = -\frac{\hbar^2}{2m}\frac{1}{r^2}\frac{\partial}{\partial r}\left(r^2\frac{\partial}{\partial r}\right) + V(r) + \frac{\hat{L}^2}{2mr^2} \qquad (5.22)$$

The final term on the right-hand side is the quantum-mechanical equivalent of the centrifugal term in the classical expression for the total energy. The only part of $\hat{H}$ which depends on θ and ϕ is this last term so it follows from this and the fact that $\hat{L}^2$ and $\hat{L}_z$ are not functions of r that $\hat{H}$ commutes with both $\hat{L}^2$ and $\hat{L}_z$, so that the total energy eigenfunctions are in this case also eigenfunctions of the total angular momentum and of one component. Energy levels corresponding to the same value of l, but different values of m are degenerate, so the angular part of the wave function may be any linear combination of the eigenfunctions corresponding to these various values of m. This is equivalent to saying that the spatial orientation of the angular momentum vector is completely unknown so that no component of it can be measured unless the spherical symmetry of the problem is broken—for example by applying a magnetic field as will be discussed in the next section.

5.3 THE EXPERIMENTAL MEASUREMENT OF ANGULAR MOMENTUM

Now that we have derived expressions for the angular momentum eigenvalues, we shall proceed to see how these agree with the results of experimental measurement. Such measurements are often made by considering the effects of magnetic fields on the motion of particles and we shall consider these in this section.

First we consider the classical motion of a particle such as an electron with charge $-e$ and mass m_e moving with angular velocity ω in an orbit of radius r. Its angular momentum $\mathbf{l}$ therefore has magnitude $m_e\omega^2 r^2$ and points in a direction perpendicular to the plane of the orbit. The circulating charge is clearly equivalent to an electric current $-e\omega/2\pi$ moving round a loop of radius πr^2. It therefore has a magnetic moment $\boldsymbol{\mu}$ in the same direction as $\mathbf{l}$ and of magnitude $-e\omega r^2$. Hence we have

$$\boldsymbol{\mu} = -\frac{e}{2m_e}\mathbf{l} \qquad (5.23)$$

We can apply Postulate 4.3 to (5.23) to obtain the quantum-mechanical operator representing a measurement of magnetic moment:

$$\hat{\boldsymbol{\mu}} = -\frac{e}{2m_e}\hat{\mathbf{L}}$$

that is,

$$\hat{\mu}_z = -\frac{e}{2m_e} \hat{L}_z \quad \text{etc.} \quad (5.24)$$

It follows that, if the z component of the magnetic moment of an atom is measured, a value for this component of the angular momentum is also obtained.

If a magnetic field **B** is applied to such a system, the energy of interaction between the magnetic moment and the field will be $-\boldsymbol{\mu} \cdot \mathbf{B}$. The corresponding quantum mechanical operator in the case where **B** is in the z direction is therefore $\Delta \hat{H}$ where

$$\Delta \hat{H} = \frac{eB}{2m_e} \hat{L}_z \quad (5.25)$$

and, if the system is in an eigenstate of $\hat{L}_z$ with eigenvalue $m\hbar$, a measurement of this interaction energy will yield the value

$$\Delta E = \mu_B Bm \quad (5.26)$$

where

$$\mu_B = \frac{e\hbar}{2m_e} \quad (5.27)$$

and is known as the *Bohr magneton*. Thus, for example, if we apply a magnetic field to an atom in a p (that is, $l = 1$) state we should expect the threefold degeneracy to be lifted and spectral lines resulting from a transition between this state and a non-degenerate s ($l = 0$) state to be split into a triplet, the angular frequency difference between neighbouring lines being equal to $(eB/2m_e)$. However, when this experiment is performed on a one-electron atom such as hydrogen, a rather different result is observed. For strong fields the spectra are consistent with the $2p$ level having been split into *four* substates instead of three and the $1s$ level, which was expected to remain single, into *two*. As the latter state possesses no angular momentum associated with the motion of the electron in the field of the nucleus, the observed magnetic moment and associated angular momentum must arise from some other cause.

We have to postulate that an electron possesses an additional *intrinsic* angular momentum in addition to the *orbital* angular momentum discussed above. In the $l = 0$ state this intrinsic angular momentum can apparently adopt two orientations, while in the $l = 1$ state the orbital and intrinsic angular momenta couple to produce the four substates observed. The details of this 'spin-orbit coupling' will be discussed in the next chapter.

The possession by the electron of two kinds of angular momentum suggests an analogy with the classical case of a planet, such as the earth, orbiting the sun: it has orbital angular momentum associated with this motion and intrinsic angular momentum associated with the fact that it is spinning about an axis. Because of this, the intrinsic angular momentum of an electron (or other particle) is usually referred to as *spin*. However, there is no evidence to suggest that the electron is literally spinning and, indeed, it turns out that such a mechanical model could account for the observed angular momentum only if the electron had a structure, parts of which would have to be moving at a speed faster than that of light, so contradicting the theory of relativity! In fact the only satisfactory theory of the origin of spin does result from a consideration of relativistic effects: Dirac showed that quantum mechanics and relativity could be made consistent only if the electron is assumed to be a point particle which possesses intrinsic angular momentum. However, Dirac's theory is well beyond the scope of this book and we shall discuss the theory of intrinsic angular momentum without considering its origins in detail. Before doing so, however, we shall describe another experiment which provides some direct evidence concerning the existence and properties of 'spin'.

The Stern–Gerlach Experiment

We have seen that the energy of an atom which possesses a magnetic moment is changed by the interaction between this magnetic moment and an applied magnetic field. If this applied field is uniform, the interaction energy will not vary from place to place and the atom will not be subject to a force, but if the field is inhomogeneous there will be a force on the atom directing it towards regions where the interaction energy is a minimum. Thus, if an atom with z component of magnetic moment μ_z is in a magnetic field directed along z whose magnitude $B(x)$ is a function of x, it will be subject to a force of magnitude F given by

$$F = \mu_z \frac{\partial B(x)}{\partial x} \tag{5.28}$$

An experiment which makes use of this effect to measure atomic magnetic moments is the *Stern–Gerlach* experiment. A beam of atoms is directed between two poles of a magnet that are shaped as shown in Fig. 5.2b in order to produce an inhomogeneous field which exerts a force on the atoms in a direction transverse to the beam. On emerging from the field, the atoms will have been deflected by an amount proportional to this component of their magnetic moment. In the absence of spin, therefore, we should expect the beam to be split into $2l + 1$ parts where l is the total orbital angular momentum quantum number, and in the particular case where l is zero we should then expect no splitting at all. However, when this experiment is performed on hydrogen (or a similar one-electron system) in its ground ($l = 0$) state the atomic beam is found to be split into two, implying that

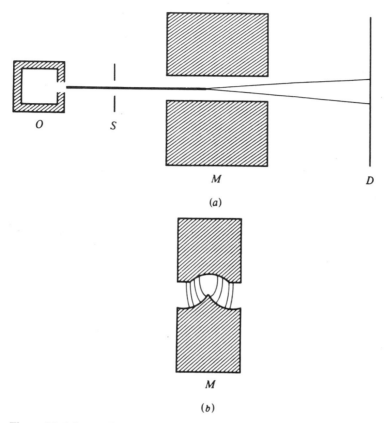

Figure 5.2 A beam of atoms leaves the oven O, passes through the slit S and is split by the inhomogeneous magnetic field M before being detected at D. The form of the field is shown in projection down the beam direction in (b).

the atom possesses a magnetic moment whose z component can adopt two opposite orientations with respect to the field. Thus the Stern–Gerlach experiment provides direct evidence for the electron possessing intrinsic angular momentum which is consistent with the spectroscopic results described above. It is clear, however, that the rules governing the quantization of spin must be different from those governing orbital angular momentum, as the latter require there to be an odd number ($2l + 1$) of eigenstates of L_z whereas an even number (2) is observed in the Stern–Gerlach experiment.

5.4 GENERAL SOLUTION TO THE EIGENVALUE PROBLEM

If we re-examine the arguments leading to the quantization of orbital angular momentum, which were set out when the equivalent problem of the energy eigenfunctions of a particle in a central field was discussed in Chapter 3, we see that it resulted from the boundary condition on the wave function requiring it to be a continuous, single-valued, integrable function of the particle position. However, spin is not associated with the motion of the particle in space and its eigenfunctions are therefore not functions of particle position, so we should not expect that the same conditions will necessarily apply in this case. On the other hand, spin is a form of angular momentum so it might be reasonable to expect the quantum-mechanical operators representing its components to obey the same algebra as do the corresponding operators in the orbital case and, in particular, we might expect the commutation relations (5.5) to (5.7) and (5.12) to be universally valid for all forms of angular momentum. We shall assume this to be the case and, using only this assumption and the basic physical condition that the total angular momentum eigenvalues must be at least as large as those of any simultaneously measured component, we shall obtain expressions for the angular-momentum eigenvalues that are valid in the spin, as well as the orbital, case.

Let the simultaneous eigenvalues of $\hat{L}^2$ and $\hat{L}_z$ be α and β respectively and let the corresponding eigenfunction be ϕ, where ϕ is not necessarily a function of particle position and a suitable representation in the case of spin will be described in the next chapter. We then have

$$\hat{L}^2\phi = \alpha\phi \quad \text{and} \quad \hat{L}_z\phi = \beta\phi \quad \text{where } \alpha \geqslant \beta^2 \quad (5.29)$$

We now define two operators $\hat{L}_+$ and $\hat{L}_-$ as

$$\hat{L}_+ = \hat{L}_x + i\hat{L}_y \quad \hat{L}_- = \hat{L}_x - i\hat{L}_y \quad (5.30)$$

These operators do not represent physical quantities (note that they are not Hermitian) but are very useful in developing the mathematical argument that follows. For reasons that will soon become clear they are known as *raising and lowering operators*, or sometimes as *ladder operators*.

We first establish some commutation relations involving $\hat{L}_+$ *and* $\hat{L}_-$. From (5.30), using (5.4) and (5.5),

$$\hat{L}_+\hat{L}_- = \hat{L}_x^2 + \hat{L}_y^2 - i[\hat{L}_x, \hat{L}_y]$$
$$= \hat{L}^2 - \hat{L}_z^2 + \hbar\hat{L}_z \quad (5.31)$$

Similarly

$$\hat{L}_-\hat{L}_+ = \hat{L}^2 - \hat{L}_z^2 - \hbar\hat{L}_z \quad (5.32)$$

Hence

$$[\hat{L}_+, \hat{L}_-] = 2\hbar\hat{L}_z \quad (5.33)$$

We also have

$$[\hat{L}_z, \hat{L}_+] = [\hat{L}_z, \hat{L}_x] + i[\hat{L}_z, \hat{L}_y]$$
$$= i\hbar(\hat{L}_y - i\hat{L}_x)$$

using (5.6) and (5.7). That is,

$$[\hat{L}_z, \hat{L}_+] = \hbar\hat{L}_+ \tag{5.34}$$

and similarly,

$$[\hat{L}_z, \hat{L}_-] = -\hbar\hat{L}_- \tag{5.35}$$

Having established these relations, which we shall refer to again shortly, we now operate on both sides of the second of these equations (5.29) with $\hat{L}_+$ giving

$$\hat{L}_+\hat{L}_z\phi = \hat{L}_+\beta\phi$$

Using (5.34) and remembering that β is a constant, this leads to

$$\hat{L}_z(\hat{L}_+\phi) = (\beta + \hbar)(\hat{L}_+\phi) \tag{5.36}$$

Similarly, using (5.29) and (5.35), we get

$$\hat{L}_z(\hat{L}_-\phi) = (\beta - \hbar)(\hat{L}_-\phi) \tag{5.37}$$

Thus, if ϕ is an eigenfunction of $\hat{L}_z$ with eigenvalue β, $(\hat{L}_+\phi)$ and $(\hat{L}_-\phi)$ are also eigenfunctions of $\hat{L}_z$ with eigenvalues $(\beta + \hbar)$ and $(\beta - \hbar)$ respectively.

We also have, from the first of the equations (5.29),

$$\hat{L}_+\hat{L}^2\phi = \alpha\hat{L}_+\phi \quad \text{and} \quad \hat{L}_-\hat{L}^2\phi = \alpha\hat{L}_-\phi \tag{5.38}$$

Now we know that $\hat{L}^2$ commutes with both $\hat{L}_x$ and $\hat{L}_y$ so it must also commute with $\hat{L}_+$ and $\hat{L}_-$; (5.38) therefore becomes

$$\hat{L}^2(\hat{L}_+\phi) = \alpha(\hat{L}_+\phi) \quad \text{and} \quad \hat{L}^2(\hat{L}_-\phi) = \alpha(\hat{L}_-\phi) \tag{5.39}$$

so the functions $(\hat{L}_+\phi)$ and $(\hat{L}_-\phi)$ as well as being eigenfunctions of $\hat{L}_z$ are also eigenfunctions of $\hat{L}^2$ with eigenvalue α. It follows that the operators $\hat{L}_+$ and $\hat{L}_-$ respectively 'raise' and 'lower' the eigenfunctions up or down the 'ladder' of eigenvalues of $\hat{L}_z$ corresponding to the same eigenvalue of $\hat{L}^2$, so accounting for the names of the operators mentioned above.†

We can now apply the condition that β^2 is less than or equal to α which implies that there must be both a maximum (say β_1) and minimum (say β_2) value of β corresponding to a particular value of α. If ϕ_1 and ϕ_2 are the corresponding eigenfunctions, we can satisfy (5.36) and (5.37) only if

$$\hat{L}_+\phi_1 = 0 \tag{5.40}$$

---

† The operators $\hat{L}_+$ and $\hat{L}_-$ are sometimes known as 'creation' and 'annihilation' operators respectively because they 'create' or 'annihilate' quanta of $\hat{L}_z$.

and

$$\hat{L}_-\phi_2 = 0 \tag{5.41}$$

We now operate on (5.40) with $\hat{L}_-$ and use (5.32) to get

$$\hat{L}_-\hat{L}_+\phi_1 = (\hat{L}^2 - \hat{L}_z^2 - \hbar\hat{L}_z)\phi_1 = 0$$

and hence, using (5.29),

$$(\alpha - \beta_1^2 - \hbar\beta_1)\phi_1 = 0$$

That is,

$$\alpha = \beta_1(\beta_1 + \hbar) \tag{5.42}$$

Similarly, from (5.41), (5.31), and (5.29), ·

$$\alpha = \beta_2(\beta_2 - \hbar) \tag{5.43}$$

It follows from (5.42) and (5.43) (remembering that β_2 must be less than β_1 by definition) that

$$\beta_2 = -\beta_1 \tag{5.44}$$

Now we saw from (5.36) and (5.37) that neighbouring values of β are separated by $\hbar$ so it follows that β_1 and β_2 are separated by an integral number (say n) of steps in $\hbar$. That is,

$$\beta_1 - \beta_2 = n\hbar \tag{5.45}$$

It follows directly from (5.44) and (5.45) that

$$\beta_1 = -\beta_2 = n\hbar/2$$

and hence, putting n equal to $2l$ and using (5.42)

$$\alpha = l(l+1)\hbar \qquad \text{where } l = n/2 \tag{5.46}$$

and

$$\beta = m\hbar \tag{5.47}$$

where m varies in integer steps between $-l$ and l, and l is either an integer or a half-integer.

We see that this result is exactly the one we have been looking for. In the case of orbital angular momentum, the extra condition that the wave function must be a well-behaved function of particle position requires l to be an integer, so (5.46) and (5.47) are then equivalent to (5.18) and (5.21). However, when we are dealing with intrinsic angular momentum this condition no longer holds and l and m can be integers or half-integers. In the particular case of electron spin, the Stern–Gerlach experiment showed that the z-component operator has two eigenstates, implying that the total spin quantum number equals one-half in this case. This is always true for the electron and also for other fundamental particles such as the proton, the neutron, and the neutrino which are all therefore known as 'spin-half' particles. Other particles exist with total-spin quantum numbers which are integers or half-integers greater than one-half, and their angular-momentum properties are also exactly as predicted by the above theory.

We mentioned above that the properties of electron spin can be derived from Dirac's relativistic quantum theory, which in fact shows that any fundamental point particle such as the electron should have total spin quantum number equal to one-half. It turns out that particles that are not ·spin-half † are either 'exchange' particles such as the photon which are not subject to Dirac's theory, or they have a structure, being composed of several even more fundamental spin-half particles: for example the alpha particle has total spin zero, but this results from a cancellation of the spins of the constituent protons and neutrons. The truly fundamental constituents of matter (apart from the photon and other 'exchange' particles) are believed to be *leptons* (which include the electron) and *quarks* and these are all spin-half particles.

In order to proceed further with our study of angular momentum, we shall have to consider the eigenfunctions as well as the eigenvalues of the angular-momentum operators. However, we have already seen that if this discussion is to include spin, these eigenfunctions cannot be expressed as functions of particle position. In the next chapter, therefore, we shall develop an alternative representation of quantum mechanics based on matrices, which will be used to describe the properties of intrinsic angular momentum.

PROBLEMS

5.1 A point mass μ is attached by a massless rigid rod of length a to a fixed point in space and is free to rotate in any direction. Find a classical expression for the energy of the system if its total angular momentum has magnitude L, and hence show that the quantum-mechanical energy levels are given by $l(l+1)\hbar^2/2ma^2$ where l is a positive integer.

5.2 The mass and rod in Prob. 5.1 are now mounted so that they can rotate only about a particular axis that is perpendicular to the rod and fixed in space. What are the energy levels now? What can be said about the components of angular momentum perpendicular to the axis of such a system when it is an energy eigenstate?

5.3 A particle has a wave function of the form $z \exp[-\alpha(x^2 + y^2 + z^2)]$ where α is a constant. Show that this function is an eigenfunction of $\hat{L}^2$ and $\hat{L}_z$ and find the corresponding eigenvalues.

---

†It is very important when referring to particles as 'spin-half', 'spin-one' etc. to remember that these labels refer to the total spin quantum number and not to the magnitude of the angular momentum which actually has the value $\sqrt{3}\hbar/2$ in the first case and $\sqrt{2}\hbar$ in the second. Similarly an electron with z component of spin equal to $\frac{1}{2}\hbar$ or $-\frac{1}{2}\hbar$ is often referred to as having 'spin up' or 'spin down' respectively even though its angular momentum vector is inclined at an angle of $\cos^{-1}(3^{-1/2}) \simeq 55°$ to the z axis.

Use the raising and lowering operators L_+ and L_- to obtain (unnormalized) expressions for all the other eigenfunctions of L_z that correspond to the same eigenvalue of $\hat{L}^2$.

Hint: Use Cartesian coordinates.

5.4 Verify directly that the spherical harmonics Y_{lm} with $l \leqslant 2$ (see Chapter 3) are eigenfunctions of $\hat{L}^2$ and $\hat{L}_z$ with the appropriate eigenvalues.

5.5 The total orbital angular momentum of the electrons in a silver atom turns out to be zero while its total spin quantum number is $\frac{1}{2}$, resulting in a magnetic moment whose z component has a magnitude of 1 Bohr magneton ($9.3 \times 10^{-24} \, \mathrm{J \, T^{-1}}$). In the original Stern–Gerlach experiment, silver atoms, each with a kinetic energy of about $3 \times 10^{-20} \, \mathrm{J}$ travelled a distance of 0.03 m through a non-uniform magnetic field of gradient $2.3 \times 10^3 \, \mathrm{T \, m^{-1}}$. Calculate the separation of the two beams 0.25 m beyond the magnet.

5.6 The wave function of a particle is known to have the form $f(r, \theta) \cos \phi$. What can be predicted about the likely outcome of a measurement of the z component of angular momentum of this system?

5.7 Show that the raising and lowering operators $\hat{L}_+$ and $\hat{L}_-$ are 'Hermitian conjugates' in the sense that $\int f L_+ g \, d\tau = \int g L_-^* f \, d\tau$.

5.8 We showed in the text that, if ϕ_m is an eigenfunction of L_z with eigenvalue $m\hbar$, then $(\hat{L}_+\phi_m)$ is an eigenfunction of $\hat{L}_z$ with eigenvalue $(m+1)\hbar$, but we did not require $(\hat{L}_+\phi_m)$ to be normalized. Making use of the result obtained in Prob. 5.7, show that the normalized eigenfunction ϕ_{m+1} is given by

$$\phi_{m+1} = [l(l+1) - m(m+1)]^{-1/2}\hbar^{-1}(\hat{L}_+\phi_m)$$

and that similarly

$$\phi_{m-1} = [l(l+1) - m(m-1)]^{-1/2}\hbar^{-1}(\hat{L}_-\phi_m)$$

5.9 Express $\hat{L}_x$ and $\hat{L}_y$ in terms of $\hat{L}_+$ and $\hat{L}_-$ and hence show that for a system in an eigenstate of $\hat{L}_z$, $\langle \hat{L}_x \rangle = \langle \hat{L}_y \rangle = 0$. Also obtain expressions for $\langle \hat{L}_x^2 \rangle$ and $\langle \hat{L}_y^2 \rangle$ (using the results obtained in Prob. 5.8 if necessary) and compare the product of these two quantities with the predictions of the uncertainty principle.

5.10 Obtain expressions for $\hat{L}_x$ and $\hat{L}_y$ in spherical polar coordinates and hence show that

$$\hat{L}_\pm = \hbar \, e^{\pm i\phi} \left(\pm \frac{\partial}{\partial \theta} + i \cot \theta \frac{\partial}{\partial \phi} \right)$$

5.11 Use the results obtained in Probs 5.8 and 5.10 to derive expressions for all the spherical harmonics with $l = 2$, given that

$$Y_{20} = \left(\frac{5}{16\pi} \right)^{1/2} (3 \cos^2 \theta - 1)$$

Compare your results with those given in Chapter 3.

SIX

ANGULAR MOMENTUM II

In the first part of this chapter we shall show how it is possible to represent dynamical variables by matrices instead of differential operators without affecting the predicted results of physically significant quantities; we shall see that, although this representation is often more complicated and cumbersome than the differential method used earlier, it has a particularly simple form when applied to problems involving angular momentum, and has the great advantage that it can be used in the case of spin where no differential operator representation exists. We shall use such spin matrices to analyse experiments designed to measure the component of spin in various directions and find that these provide an important and illustrative example of the quantum theory of measurement.

In the second part of this chapter we shall discuss the problem of the addition of different angular momenta (such as the orbital and spin angular momenta of an electron in an atom) and illustrate the results by discussing the effects of spin-orbit coupling and applied magnetic fields on the spectra of one-electron atoms.

6.1 MATRIX REPRESENTATIONS

We first consider the case of a dynamical variable represented by the Hermitian operator $\hat{Q}$. If one of its eigenvalues is q and the corresponding eigenfunction is ψ then

$$\hat{Q}\psi = q\psi \tag{6.1}$$

We now expand ψ in terms of some complete orthonormal set of functions ϕ_n, which are not necessarily eigenfunctions of $\hat{Q}$:

$$\psi = \sum_n a_n \phi_n \qquad (6.2)$$

As in Chapter 4, we assume that the ϕ_n form a discrete set, but the extension to the continuous case is reasonably straightforward. Substituting from (6.2) into (6.1) we get

$$\sum_n a_n \hat{Q} \phi_n = q \sum_n a_n \phi_n \qquad (6.3)$$

We now multiply both sides of (6.3) by the complex conjugate of one of the functions—say ϕ_m^*—and integrate over all space. This leads to

$$\sum_n Q_{mn} a_n = q a_m \qquad (6.4)$$

where Q_{mn} are defined by

$$Q_{mn} = \int \phi_m^* \hat{Q} \phi_n \, d\tau \qquad (6.5)$$

and we have used the orthogonality property in deriving the right-hand side of (6.4). Equation (6.4) is true for all values of m so we can write it in matrix form:

$$\begin{bmatrix} Q_{11} & Q_{12} & \cdots \\ Q_{21} & Q_{22} & \cdots \\ \cdot & \cdot & \cdots \\ \cdot & \cdot & \cdots \\ \cdot & \cdot & \cdots \end{bmatrix} \begin{bmatrix} a_1 \\ a_2 \\ \cdot \\ \cdot \\ \cdot \end{bmatrix} = q \begin{bmatrix} a_1 \\ a_2 \\ \cdot \\ \cdot \\ \cdot \end{bmatrix} \qquad (6.6)$$

where it is clear from (6.5) and the fact that $\hat{Q}$ is a Hermitian operator, that $Q_{mn}^* = Q_{nm}$. A matrix whose elements obey this condition is known as a *Hermitian matrix*, and Eq. (6.6) is a matrix eigenvalue equation; such an equation has solutions only for the particular values of q that satisfy the determinantal condition

$$\begin{vmatrix} Q_{11} - q & Q_{12} & \cdots \\ Q_{21} & Q_{22} - q & \cdots \\ \cdot & \cdot & \cdots \\ \cdot & \cdot & \cdots \end{vmatrix} = 0 \qquad (6.7)$$

There are as many values of q which are solutions to this equation as there are rows (or columns) in the determinant, but as we have already defined q as one of the eigenvalues of the original operator $\hat{Q}$, it follows that the eigenvalues of the matrix equation (6.6) are the same as those of the original differential equation (6.1) and these two equations are therefore formally equivalent. Equation (6.6) can therefore be written in the form

$$[Q][a] = q[a] \tag{6.8}$$

where $[Q]$ is the matrix whose elements are Q_{mn} and $[a]$ is the column vector whose elements are a_n. In the same way that ψ is an eigenfunction of the Hermitian operator $\hat{Q}$, the column vector $[a]$ is an eigenvector of the Hermitian matrix $[Q]$.

The properties of Hermitian matrices are very similar to those of Hermitian operators discussed in Chapter 4. For example, we can show that the eigenvalues of Hermitian matrices are real and the eigenvectors corresponding to different eigenvalues are orthogonal by an argument very similar to that leading to Eqs (4.15) and (4.25): let $[a]_1$, and $[a]_2$ be eigenvectors of $[Q]$ with eigenvalues q_1 and q_2 respectively. That is,

$$[Q][a]_1 = q_1[a]_1 \tag{6.9}$$

and

$$[Q][a]_2 = q_2[a]_2 \tag{6.10}$$

If $[Q]$ is Hermitian in the sense described above, and if $[\tilde{a}]_1$ and $[\tilde{a}]_2$ are row vectors whose elements are the complex conjugates of those of $[a]_1$ and $[a]_2$ respectively, it follows from the standard rules of matrix multiplication that

$$[\tilde{a}]_1[Q] = q_1^*[\tilde{a}]_1 \tag{6.11}$$

and

$$[\tilde{a}]_2[Q] = q_2^*[\tilde{a}]_2 \tag{6.12}$$

If we now pre-multiply (6.9) by $[\tilde{a}]_1$ and post-multiply (6.11) by $[a]_1$, the left-hand sides of the resulting equations are equal, so we can equate the right-hand sides to get

$$q_1[\tilde{a}]_1[a]_1 = q_1^*[\tilde{a}]_1[a]_1$$

that is,

$$q_1 = q_1^* \tag{6.13}$$

and the eigenvalue is real as required. To prove orthogonality we pre-multiply (6.10) by $[\tilde{a}_1]$ and post-multiply (6.11) by $[a_2]$; the left-hand sides are again equal and equating the right-hand sides gives

$$q_2[\tilde{a}]_1[a]_2 = q_1^*[\tilde{a}]_1[a]_2$$

and therefore, assuming $q_1 \neq q_2$

$$[\tilde{a}]_1 [a]_2 = 0 \tag{6.14}$$

By orthogonality of eigenvectors we therefore mean that the scalar product of one with the row vector formed from the complex conjugate of the other is zero. Clearly we can also ensure normalization, in the sense that $[\tilde{a}]_1 [a]_1 = 1$, by multiplying all the elements of the eigenvector by an appropriate constant. The other properties of Hermitian operators set out in Chapter 4 can be converted to matrix form in a similar manner, the main difference being that where integrals of products of operators and functions appear in the former case, they are replaced by products of matrices and vectors in the latter.

We see from the above that the physical properties of a quantum-mechanical system can in principle be derived using appropriate matrix equations instead of the differential operator formalism. The postulates and algebra contained in Chapter 4 can all be expressed directly in matrix terms: the dynamical variables are represented by Hermitian matrices and the state of the system is described by a 'state vector' that is identical with the appropriate eigenvector immediately after a measurement whose result was equal to the corresponding eigenvalue. Moreover, provided we could postulate appropriate matrices to represent the dynamical variables, the whole procedure could be carried out without ever referring to the original operators or eigenfunctions.

However, the matrix method suffers from one major disadvantage. This is that, because most quantum systems have an infinite number of eigenstates, the dynamical variables, including position and momentum, usually have to be represented by matrices of infinite order. Techniques for obtaining the eigenvalues in some such cases have been developed and, indeed, Heisenberg developed a form of quantum mechanics based on matrices (sometimes known as 'matrix mechanics') and used it to solve the energy eigenvalue equations for the simple harmonic oscillator and the hydrogen atom before 'wave mechanics' had been invented by Schrödinger. Nevertheless, matrices of infinite order are generally difficult to handle and the solution of the corresponding differential equations is nearly always easier. However, this particular problem does not arise when the number of eigenvalues is finite, which accounts for the usefulness of matrix methods when studying angular momentum. Provided the total angular momentum has a fixed value (determined by the quantum number l) only the $2l + 1$ states with different values of the quantum number m have to be considered and the angular momentum operators can be represented by matrices of order $2l + 1$. This simplification is particularly useful in the case of spin, as for spin-half particles, $l = \frac{1}{2}$ and therefore 2×2 matrices can be used; moreover we have seen that the spin components cannot be represented by differential operators so matrix methods are particularly useful in this case.

6.2 PAULI SPIN MATRICES

A suitable set of 2×2 matrices which can be used to represent the angular-momentum components of a spin-half particle were first discovered by W. Pauli and are known as *Pauli spin matrices*. These have the form:

$$[\sigma_x] = \begin{bmatrix} 0 & 1 \\ 1 & 0 \end{bmatrix} \quad [\sigma_y] = \begin{bmatrix} 0 & -i \\ i & 0 \end{bmatrix} \quad [\sigma_z] = \begin{bmatrix} 1 & 0 \\ 0 & -1 \end{bmatrix} \quad (6.15)$$

and the operators $\hat{S}_x$, $\hat{S}_y$, and $\hat{S}_z$ representing the three spin components are expressed in the matrix form as:

$$\hat{S}_x = \tfrac{1}{2}\hbar[\sigma_x] \quad \text{etc.} \quad (6.16)$$

To see that these matrices do indeed form a suitable representation for spin, we first check whether they obey the correct commutation relations.

$$[\hat{S}_x, \hat{S}_y] = \tfrac{1}{4}\hbar^2 \left(\begin{bmatrix} 0 & 1 \\ 1 & 0 \end{bmatrix} \begin{bmatrix} 0 & -i \\ i & 0 \end{bmatrix} - \begin{bmatrix} 0 & -i \\ i & 0 \end{bmatrix} \begin{bmatrix} 0 & 1 \\ 1 & 0 \end{bmatrix} \right)$$

$$= \tfrac{1}{4}\hbar^2 \left(\begin{bmatrix} i & 0 \\ 0 & -i \end{bmatrix} - \begin{bmatrix} -i & 0 \\ 0 & i \end{bmatrix} \right)$$

$$= i\hbar\tfrac{1}{2}\hbar \begin{bmatrix} 1 & 0 \\ 0 & -1 \end{bmatrix}$$

$$= i\hbar S_z \quad (6.17)$$

This result agrees with that derived earlier (5.5) and the other commutation relations can be similarly verified. Turning now to the eigenvalues, we see that those of $\hat{S}_z$ are simply $\tfrac{1}{2}\hbar$ times the eigenvalues of $[\sigma_z]$ which are in turn obtained from the 2×2 equivalent of (6.7):

$$\begin{vmatrix} 1 - \lambda & 0 \\ 0 & -1 - \lambda \end{vmatrix} = 0 \quad (6.18)$$

Hence $\lambda = \pm 1$ and the eigenvalues of $\hat{S}_z$ are $\pm \tfrac{1}{2}\hbar$ as expected. The corresponding eigenvectors are obtained from the eigenvalue equation

$$\tfrac{1}{2}\hbar \begin{bmatrix} 1 & 0 \\ 0 & -1 \end{bmatrix} \begin{bmatrix} a_1 \\ a_2 \end{bmatrix} = \pm \tfrac{1}{2}\hbar \begin{bmatrix} a_1 \\ a_2 \end{bmatrix} \quad (6.19)$$

which, along with the normalization condition, leads to the expressions $\begin{bmatrix} 1 \\ 0 \end{bmatrix}$

and $\begin{bmatrix} 0 \\ 1 \end{bmatrix}$ for the eigenvectors corresponding to the eigenvalues $\tfrac{1}{2}\hbar$ and $-\tfrac{1}{2}\hbar$ respectively.

Eigenvalues and eigenvectors can similarly be found for $\hat{S}_x$ and $\hat{S}_y$. In each case the eigenvalues are equal to $\pm\frac{1}{2}\hbar$ and the corresponding eigenvectors are shown in Table 6.1. The reader should verify that each eigenvector is normalized and that each pair is orthogonal.

As a further test of the correctness of this representation, we consider the square of the total angular momentum whose operator is given by

$$\hat{S}^2 = \hat{S}_x^2 + \hat{S}_y^2 + \hat{S}_z^2$$

$$= \tfrac{3}{4}\hbar^2 \begin{bmatrix} 1 & 0 \\ 0 & 1 \end{bmatrix} \tag{6.20}$$

using (6.15) and (6.16). Clearly $\hat{S}^2$ commutes with $\hat{S}_x$, $\hat{S}_y$, and $\hat{S}_z$ as expected; moreover it has a pair of degenerate eigenstates whose eigenvalue is $\tfrac{3}{4}\hbar^2$ and whose eigenvectors are in common with those of any of the components. We conclude therefore that the Pauli spin matrix representation for the angular momentum components of a spin-half particle produces the same results as were previously obtained by other methods and we can proceed with confidence to use this representation to obtain further information about the properties of spin.

Table 6.1 The eigenvalues and eigenvectors of the matrices representing the angular momentum components of a spin-half particle

Spin component	Eigenvalue	Eigenvector
$\hat{S}_x$	$\tfrac{1}{2}\hbar$	$\alpha_x = \dfrac{1}{\sqrt{2}}\begin{bmatrix} 1 \\ 1 \end{bmatrix}$
$\hat{S}_x$	$-\tfrac{1}{2}\hbar$	$\beta_x \equiv \dfrac{1}{\sqrt{2}}\begin{bmatrix} -1 \\ 1 \end{bmatrix}$
$\hat{S}_y$	$\tfrac{1}{2}\hbar$	$\alpha_y \equiv \dfrac{1}{\sqrt{2}}\begin{bmatrix} 1 \\ i \end{bmatrix}$
$\hat{S}_y$	$-\tfrac{1}{2}\hbar$	$\beta_y \equiv \dfrac{1}{\sqrt{2}}\begin{bmatrix} 1 \\ -i \end{bmatrix}$
$\hat{S}_z$	$\tfrac{1}{2}\hbar$	$\alpha_z \equiv \begin{bmatrix} 1 \\ 0 \end{bmatrix}$
$\hat{S}_z$	$-\tfrac{1}{2}\hbar$	$\beta_z \equiv \begin{bmatrix} 0 \\ 1 \end{bmatrix}$

6.3 SPIN AND THE QUANTUM THEORY OF MEASUREMENT

The measurement of spin provides a very clear illustration of the quantum theory of measurement. Consider a beam of spin-half particles travelling along the y axis, as in Fig. 6.1, towards a Stern–Gerlach apparatus oriented to measure the z component of spin (such an apparatus will be denoted by SGZ below). We assume that the number of particles in each beam is small enough for there to be only one in the Stern–Gerlach apparatus at any one time; we can therefore treat each particle independently and ignore interactions between them. Two beams of particles will emerge, one with z component equal to $-\frac{1}{2}\hbar$, which we block off with a suitable stop, and the other with z component $\frac{1}{2}\hbar$, which we allow to continue. This measurement has therefore defined the state of the system and the state vector of the particles which proceed is known to be the eigenvector α_z of $\hat{S}_z$, corresponding to an eigenvalue of $\frac{1}{2}\hbar$ (cf. Table 6.1). If we were to pass this beam directly through another SGZ apparatus, we should expect all the particles to emerge in the channel corresponding to $S_z = \frac{1}{2}\hbar$ and none to emerge in the other channel, and indeed this prediction is confirmed experimentally.

We now consider the case where the particles that emerge from the first SGZ with positive z component pass into a similar apparatus oriented now to measure the x component of spin (that is an SGX) as in Fig. 6.1. Each particle must emerge from the SGX with S_x equal to either $\frac{1}{2}\hbar$ or $-\frac{1}{2}\hbar$, but, as the initial state vector is not an eigenvector of $\hat{S}_x$, we cannot predict which result will actually be obtained. However, we can use the quantum theory of measurement to predict the relative probabilities of the possible outcomes: we expand the initial state vector as a linear combination of the eigenvectors of $\hat{S}_x$ and the probabilities are then given by the squares of the appropriate expansion coefficients. Thus, referring to Table 6.1,

$$\alpha_z = c_+\alpha_x + c_-\beta_x \qquad (6.21)$$

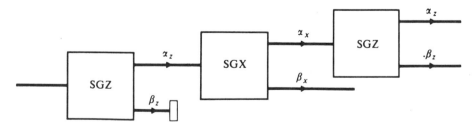

Figure 6.1 Successive measurements of the angular momentum components of spin-half particles. The boxes represent sets of Stern–Gerlach apparatus which direct a particle into the upper or lower output channel depending on whether the appropriate spin component is positive or negative.

where c_+ and c_- are constants. That is,

$$\begin{bmatrix} 1 \\ 0 \end{bmatrix} = c_+ \frac{1}{\sqrt{2}} \begin{bmatrix} 1 \\ 1 \end{bmatrix} + c_- \frac{1}{\sqrt{2}} \begin{bmatrix} 1 \\ -1 \end{bmatrix} \qquad (6.22)$$

from which it follows that $c_+ = c_- = 2^{-1/2}$ and therefore

$$\alpha_z = \frac{1}{\sqrt{2}} (\alpha_x + \beta_x) \qquad (6.23)$$

Thus the two possible results of a measurement of the z component of spin have equal probability and, although we cannot predict the result of the measurement on any particular particle, we know that if the experiment is repeated a large number of times, half the particles will emerge in each channel. However, it is important to realize that this does not mean that, before the second measurement, the particles were randomly distributed between the eigenstates of $\hat{S}_x$ with half having each eigenvalue. This is because, if this had been the case, they could not all have passed through the same channel of the first (SGZ) apparatus as their state vectors would not have been eigenvectors of $\hat{S}_z$. To emphasize this point, consider one of the beams of particles emerging from SGX to be passed through a second SGZ (see Fig. 6.1 again); a similar argument to the above shows that we shall again be unable to predict in which channel a particular particle will emerge, although the relative probabilities of both possible results are again equal.

We see therefore that the act of measuring one component of spin has destroyed any knowledge we previously had about the value of another. In fact the quantum theory of measurement (at least as it is conventionally interpreted) states that while a particle is in an eigenstate of $\hat{S}_z$, it is meaningless to think of it as having a value for the x-component of its spin, as the latter can be measured only by changing the state of the system to be an eigenstate of $\hat{S}_x$. Moreover, as we have seen, the result of a measurement of the x component is completely unpredictable if the particle is in an eigenstate of $\hat{S}_z$, and this indeterminacy is to be thought of as an intrinsic property of such a quantum system. Some scientists have found these features of quantum mechanics very unsatisfactory for a number of reasons and attempts have been made to develop other theories, known as 'hidden variable theories', that avoid the problems of indeterminacy and the like. We shall discuss these in some detail in Chapter 11 where we shall show that, when such theories predict results different from those of quantum mechanics, experiment has always favoured the latter. In that chapter we shall also discuss the measurement problem in more depth, again making use of the measurement of spin components as an illustrative example. These discussions will involve the consideration of the measurement of two spin components that are not necessarily at right angles to each other and, as this also provides another example of the application of Pauli spin matrices, we shall derive some relevant results for this case at this point.

Consider a beam of particles, travelling along the y axis, that has passed through the positive channel of an SGZ apparatus and is then directed into a second Stern–Gerlach apparatus (SGϕ) oriented to measure the spin component in the xz plane at an angle ϕ to z. The operator $\hat{S}_\phi$, representing the component of angular momentum in this direction, is given by analogy with the classical expression as

$$\hat{S}_\phi = \hat{S}_z \cos \phi + \hat{S}_x \sin \phi$$

$$= \tfrac{1}{2}\hbar \begin{bmatrix} \cos \phi & \sin \phi \\ \sin \phi & -\cos \phi \end{bmatrix} \tag{6.24}$$

using (6.15). The eigenvalues of this matrix are easily shown to be $\pm\tfrac{1}{2}\hbar$ as expected and if the eigenvector in the positive case is $\begin{bmatrix} a_1 \\ a_2 \end{bmatrix}$, then

$$\tfrac{1}{2}\hbar(a_1 \cos \phi + a_2 \sin \phi) = \tfrac{1}{2}\hbar a_1$$

and therefore

$$\frac{a_1}{a_2} = \frac{\sin \phi}{1 - \cos \phi} = \frac{\cos(\phi/2)}{\sin(\phi/2)} \tag{6.25}$$

The normalized eigenvector is therefore $\begin{bmatrix} \cos(\phi/2) \\ \sin(\phi/2) \end{bmatrix}$ and a similar argument leads to the expression $\begin{bmatrix} -\sin(\phi/2) \\ \cos(\phi/2) \end{bmatrix}$ in the case where the eigenvalue equals $-\tfrac{1}{2}\hbar$. We note that when $\phi = 0$, the eigenvectors are the same as those determined above for $\hat{S}_z$; and when $\phi = \pi/2$ they are identical to those of $\hat{S}_x$ (cf. Table 6.1); this is to be expected as the ϕ direction is parallel to z and x respectively in these cases.

Suppose now that a beam of particles whose z component is known to be $\tfrac{1}{2}\hbar$ (presumably because they have previously passed through the appropriate channel of an SGZ apparatus) enters an SGϕ apparatus. The probabilities of obtaining a positive or negative result from such a measurement are $|c_+|^2$ and $|c_-|^2$ respectively, where c_+ and c_- are now the coefficients of the SGϕ eigenvectors in the expansion of the initial state vector. Thus

$$\begin{bmatrix} 1 \\ 0 \end{bmatrix} = c_+ \begin{bmatrix} \cos(\phi/2) \\ \sin(\phi/2) \end{bmatrix} + c_- \begin{bmatrix} -\sin(\phi/2) \\ \cos(\phi/2) \end{bmatrix} \tag{6.26}$$

from which it follows that

$$c_+ = \cos(\phi/2) \quad \text{and} \quad c_- = -\sin(\phi/2) \tag{6.27}$$

and the probabilities of a positive or negative result are therefore $\cos^2(\phi/2)$ and $\sin^2(\phi/2)$ respectively. We note that when $\phi = 0$ the result is certain to be positive because in this case the second apparatus is identical to the first, and that when $\phi = \pi/2$ (corresponding to a measurement of the x component) the probabilities are both equal to one-half in agreement with the earlier result. Experiments have been performed to carry out measurements such as those described above, and in each case they produce results in agreement with those predicted by the quantum theory of measurement. We shall discuss such experiments in more detail in Chapter 11 where we shall find that the confirmation of results such as (6.27) is an important test of the correctness of quantum mechanics when compared with 'hidden variable' theories.

6.4 THE ADDITION OF ANGULAR MOMENTA

We saw in the last chapter that an electron in an atom possesses two kinds of angular momenta: orbital angular momentum associated with its motion, and intrinsic angular momentum or 'spin'. Many of the detailed proportion of atoms can be understood by considering the ways in which the energy levels are affected by interactions between these quantities, which effect is known as *spin-orbit coupling*. Similar interactions are also important in the study of the structure of the nucleus, where the spins and orbital angular momenta of the constituent nucleons are coupled, and of that of fundamental particles where the relevant quantities are the spins of the individual quarks. To understand these couplings we first consider the general problem of the addition of angular momenta and then use the results to discuss spin-orbit coupling and the Zeeman effect in one-electron atoms later in this chapter.

We consider two angular momenta whose squared magnitudes are represented by the operators $\hat{L}_1^2$ and $\hat{L}_2^2$ and whose z components are represented by $\hat{L}_{z1}$ and $\hat{L}_{z2}$. We assume that the two angular momenta can be measured compatibly so all four operators must commute. That is,

$$[\hat{L}_1^2, \hat{L}_{z1}] = [\hat{L}_2^2, \hat{L}_{z2}] = [\hat{L}_1^2, \hat{L}_2^2] = [\hat{L}_1^2, \hat{L}_{z2}]$$
$$= [\hat{L}_{z1}, \hat{L}_2^2] = [\hat{L}_{z1}, \hat{L}_{z2}] = 0 \qquad (6.28)$$

The four operators therefore have a common set of eigenfunctions (or eigenvectors if a matrix representation is being used) which will be specified by the quantum numbers l_1, m_1, l_2, and m_2 where these are all either integers or half-integers and where $-l_1 \leqslant m_1 \leqslant l_1$ and $-l_2 \leqslant m_2 \leqslant l_2$, as discussed in the previous chapter. The eigenfunction $\chi_{l_1 m_1 l_2 m_2}$ is therefore simply a product of the individual eigenfunctions. That is,

$$\chi_{l_1 m_1 l_2 m_2} = Y_{l_1 m_1} Y_{l_2 m_2} \qquad (6.29)$$

where we have assumed here that $\hat{L}_1$ and $\hat{L}_2$ both represent orbital angular momenta, but similar expressions can be written in matrix notation in the spin case. The fact that the expressions (6.29) are eigenfunctions of all four operators follows immediately on substitution into the appropriate eigenvalue equations.

The above discussion has assumed that the values of the total and z component of both angular momenta are known. Although this condition is sometimes fulfilled at least approximately (as in the strong-field Zeeman effect to be discussed below) situations often arise in which the values of the individual z components are unknown. The quantities measured in this case are, typically, the individual magnitudes whose squares are represented by $\hat{L}_1^2$ and $\hat{L}_2^2$ as before, and the squared magnitude and z component of the total which are represented by the operators $\hat{L}^2$ and $\hat{L}_z$ respectively where

$$\hat{L}^2 = |\hat{\mathbf{L}}_1 + \hat{\mathbf{L}}_2|^2 = \hat{L}_1^2 + \hat{L}_2^2 + 2\hat{\mathbf{L}}_1 \cdot \hat{\mathbf{L}}_2 \tag{6.30}$$

and

$$\hat{L}_z = \hat{L}_{z1} + \hat{L}_{z2} \tag{6.31}$$

the scalar product $\hat{\mathbf{L}}_1 \cdot \hat{\mathbf{L}}_2$ being defined by analogy with the standard classical expression as

$$\hat{\mathbf{L}}_1 \cdot \hat{\mathbf{L}}_2 = \hat{L}_{x1}\hat{L}_{x2} + \hat{L}_{y1}\hat{L}_{y2} + \hat{L}_{z1}\hat{L}_{z2} \tag{6.32}$$

Clearly $\hat{L}^2$ commutes with $\hat{L}_1^2$ and $\hat{L}_2^2$ and also with $\hat{L}_z$, but not with $\hat{L}_{z1}$ or $\hat{L}_{z2}$ (these relations can be checked by substituting from (6.30), (6.31) and (6.32) and using the expressions $[\hat{L}_{x1}, \hat{L}_{y1}] = i\hbar\hat{L}_{z1}$, etc.). It follows from the general discussion of degeneracy at the end of Chapter 4 that the operators L^2, L_z, L_1^2 and L_2^2 must possess a common set of eigenfunctions. We now explain how to obtain these along with the eigenvalues associated with these operators.

The eigenvalues of the square of the magnitude of any angular momentum vector and its corresponding z component were determined in the previous chapter, using a very general argument based only on the commutation relations between the operators representing the components. The results of this can therefore be applied to the present case and the eigenvalues of $\hat{L}^2$, $\hat{L}_z$, $\hat{L}_1^2$, and $\hat{L}_2^2$ are therefore $l(l+1)\hbar^2$, $m\hbar$, $l_1(l_1+1)\hbar^2$ and $l_2(l_2+1)\hbar^2$ respectively where l, l_1, l_2 and m are integers or half-integers and $-l \leqslant m \leqslant l$. The common eigenfunctions can therefore be denoted by these four quantum numbers and written as $\phi_{lml_1l_2}$. To find expressions for these we first of all consider again the simple products $Y_{l_1m_1}Y_{l_2m_2}$ which were shown above to be eigenfunctions of the operators $\hat{L}_1^2$, $\hat{L}_2^2$, $\hat{L}_{z1}$ and $\hat{L}_{z2}$. (We note again that, although we are using the notation for spherical harmonics, equivalent expressions could be written in the matrix representation.) If we operate on these products with $\hat{L}_z$ we obtain, using (6.31),

$$\hat{L}_z Y_{l_1m_1}Y_{l_2m_2} = (\hat{L}_{z1} + \hat{L}_{z2})Y_{l_1m_1}Y_{l_2m_2}$$

$$= (m_1 + m_2)\hbar Y_{l_1m_1}Y_{l_2m_2} \tag{6.33}$$

It therefore follows that the products are also eigenfunctions of $\hat{L}_z$ with eigenvalues $m\hbar$ where

$$m = m_1 + m_2 \tag{6.34}$$

and that all such products that have the same values of l_1, l_2 and m form a degenerate set with respect to the operators $\hat{L}_1^2$, $\hat{L}_2^2$ and $\hat{L}_z$. Physically this means that the total z component of angular momentum must be the sum of the two individual z components, and any allowed orientation of $\mathbf{L}_1$ and $\mathbf{L}_2$ that satisfies this condition and has the correct value for the total angular momentum is a possible state of the system. Referring to the general discussions of degenerate systems at the end of Chapter 4, we see that any linear combination of these degenerate eigenfunctions is also an eigenfunction with the same eigenvalue, so we should be able to express the eigenfunctions we are looking for as linear combinations of the members of this degenerate set, with the expansion coefficients chosen so that they are also eigenfunctions of $\hat{L}^2$. To do this we first consider the case where m_1 and m_2 have their maximum possible values which are l_1 and l_2 respectively. There is only one product, $Y_{l_1 l_1} Y_{l_2 l_2}$, corresponding to these quantum numbers and this must therefore be the required common eigenfunction in this case. Moreover, it follows from (6.34) that the corresponding value of m is equal to $(l_1 + l_2)$ and, as this is the maximum possible value of m, it must also be equal to l. We therefore have

$$\phi_{l_1 + l_2, l_1 + l_2, l_1, l_2} = Y_{l_1 l_1} Y_{l_2 l_2} \tag{6.35}$$

If we now consider the second highest value of m (that is, $l_1 + l_2 - 1$) there are two corresponding products, $Y_{l_1, l_1 - 1} Y_{l_2 l_2}$ and $Y_{l_1 l_1} Y_{l_2, l_2 - 1}$, and therefore two independent linear combinations can be constructed from these. One of them corresponds to the same value of l as before (that is, $l_1 + l_2$) while the other must have l equal to $l_1 + l_2 - 1$. We can therefore write

$$\phi_{l_1 + l_2, l_1 + l_2 - 1, l_1, l_2} = A Y_{l_1 l_1} Y_{l_2, l_2 - 1} + B Y_{l_1, l_1 - 1} Y_{l_2 l_2} \tag{6.36}$$

and

$$\phi_{l_1 + l_2 - 1, l_1 + l_2 - 1, l_1, l_2} = C Y_{l_1 l_1} Y_{l_2, l_2 - 1} + D Y_{l_1, l_1 - 1} Y_{l_2 l_2} \tag{6.37}$$

where the constants A, B, C, and D are known as Clebsch–Gordan coefficients and remain to be determined. This process can be continued with the value of m being reduced by 1 at each stage until its minimum value $(-l_1 - l_2)$ is reached when there is again only one product, corresponding to $m_1 = -l_1$ and $m_2 = -l_2$. Table 6.2 shows the possible values of m_1, m_2 and l corresponding to different values of m in the case where $l_1 = 2$ and $l_2 = 1$. We see that the values of l range from $l_1 + l_2$ (that is, 3) down to $|l_1 - l_2|$ (that is, 1) and that there are $2l + 1$ values of m corresponding to a given value of l as expected. Physically the maximum and minimum values of l correspond to the two angular momenta being as nearly parallel or anti-parallel respectively as is allowed by the quantization conditions.

Table 6.2. The values of the quantum numbers m_1, m_2, and l corresponding to the different values of m for the case where $l_1 = 2$ and $l_2 = 1$

(m_1, m_2)	m	l
$(2, 1)$	3	3
$(2, 0), (1, 1)$	2	3, 2
$(2, -1), (1, 0), (0, 1)$	1	3, 2, 1
$(1, -1), (0, 0), (-1, 1)$	0	3, 2, 1
$(0, -1), -1, 0), (-2, 1)$	-1	3, 2, 1
$(-2, 0), (-1, -1)$	-2	3, 2
$(-2, -1)$	-3	3

Eigenfunctions of the total angular momentum with given values of l and m are constructed as linear combinations of products of the angular momentum eigenfunctions denoted by the values of (m_1, m_2) in the appropriate row of the above table.

We shall now obtain values for the Clebsch–Gordan coefficients in the simplest possible case, which is that where $l_1 = l_2 = \frac{1}{2}$. This corresponds to the coupling of the spins of two spin-half particles such as the two electrons in the helium atom which will be discussed further in Chapter 10. The highest value of m is now 1 and we have (cf. (6.35))

$$\phi_{1\,1\,1/2\,1/2} = \alpha_1 \alpha_2 \tag{6.38}$$

where α_1 and α_2 represent eigenstates where the first and second particles respectively have positive z components of spin. We can obtain the state where $l = 1$ and $m = 0$ from (6.38) using the ladder operators introduced at the end of Chapter 5 (p. 101). We first note that it follows directly from the definitions of these operators (5.30) that the ladder operators corresponding to states of the combined angular momentum are just sums of these corresponding to the angular momenta of the individual electrons. That is

$$\left. \begin{aligned} \hat{L}_+ &= \hat{L}_{+1} + \hat{L}_{+2} \\[6pt] \hat{L}_- &= \hat{L}_{-1} + \hat{L}_{-2} \end{aligned} \right\} \tag{6.39}$$

and

Operating on (6.38) with $\hat{L}_-$ and using the second of (6.39) we get

$$\hat{L}_- \phi_{1\,1\,1/2\,1/2} = \alpha_1 \hat{L}_{-2} \alpha_2 + \alpha_2 \hat{L}_{-1} \alpha_1 \tag{6.40}$$

Hence, using the normalization results set out in Problem 5.8,

$$\hbar(1 \times 2 - 1 \times 0)^{1/2} \phi_{10\,1/2\,1/2} = \hbar[\tfrac{1}{2} \times \tfrac{3}{2} - \tfrac{1}{2} \times (-\tfrac{1}{2})](\alpha_1\beta_2 + \alpha_2\beta_1)$$

i.e.

$$\phi_{10\,1/2\,1/2} = 2^{-1/2}(\alpha_1\beta_2 + \alpha_2\beta_1) \tag{6.41}$$

Operating on this state with L_- gives

$$\phi_{1-1\,1/2\,1/2} = \beta_1\beta_2 \tag{6.42}$$

One more state, that with $l = 0$ and $m = 0$, remains to be determined. We first write this in the general form (6.37) as

$$\phi_{00\,1/2\,1/2} = C\alpha_1\beta_2 + D\alpha_2\beta_1 \tag{6.43}$$

This must be orthogonal to $\phi_{10\,1/2\,1/2}$ defined in (6.41) above so, using this and remembering that α and β are an orthogonal pair, we get

$$2^{-1/2}C + 2^{-1/2}D = 0 \tag{6.44}$$

Hence $C = -D = 2^{-1/2}$ if the combined state is also normalized, and we get

$$\phi_{00\,1/2\,1/2} = 2^{-1/2}(\alpha_1\beta_2 - \alpha_2\beta_1) \tag{6.45}$$

Procedures similar to the above can be applied to the general case. The set of functions of maximum l are obtained by successive application of the lowering operator starting with the state of maximum m. A function with next-lowest l and m is obtained using orthogonality in the manner just described and others with the same l by further applications of the lowering operators. This process is then repeated until the full set of eigenvalues is obtained. Details of these procedures are given in more advanced texts (e.g., A. R. Edmonds, *Angular Momentum in Quantum Mechanics*, Princeton University Press, New Jersey, 1957). However, considerable understanding of the problem of combining angular momenta can be achieved simply by classifying the quantum numbers of the states in the manner described earlier, and we shall use this procedure to study the problem of spin-orbit coupling in the next section.

6.5 SPIN-ORBIT COUPLING

In the last chapter we showed that, classically, an electron with orbital angular momentum **l** has a magnetic moment μ_l given by

$$\mu_l = -\frac{e}{2m_e}\mathbf{l} \tag{6.46}$$

and that the corresponding quantum mechanical operators $\hat{\mu}_l$ and $\hat{\mathbf{L}}$ are similarly related. There is a similar magnetic moment μ_s associated with electron spin which is proportional to the spin angular momentum ($\mathbf{s}$), but the constant of proportionality in this case cannot be calculated from simple theory and must be obtained from experiment or from relativistic quantum theory; its value turns out to be almost exactly twice that in the orbital case and we therefore have

$$\mu_s = -\frac{e}{m_e}\mathbf{s} \qquad (6.47)$$

The interaction between the orbital and spin magnetic moments introduces a small term into the energy of a one-electron atom which we have not previously considered; this effect is known as *spin-orbit coupling*. To obtain an expression for the quantum-mechanical operator representing the spin-orbit energy, we follow the same procedure as previously and calculate an expression for the corresponding classical quantity, which we then quantize following the procedure set out in Chapter 4.

Classically, if a particle moves with velocity $\mathbf{v}$ in an electric field $\mathbf{F}(\mathbf{r})$ then, within the frame of reference of the particle, a magnetic field $\mathbf{B}$ is created which is given by

$$\mathbf{B} = -\mathbf{v} \times \mathbf{F}/c^2 \qquad (6.48)$$

where c is the velocity of light. (This expression is derived in most standard textbooks on electromagnetic theory.) If the potential V associated with the field $\mathbf{F}$ is spherically symmetric (as in a one-electron atom) we have

$$\mathbf{F} = -\frac{\partial V}{\partial r}\frac{\mathbf{r}}{r} \qquad (6.49)$$

and hence

$$\mathbf{B} = \frac{\mathbf{v} \times \mathbf{r}}{rc^2}\frac{\partial V}{\partial r}$$

$$= -\frac{1}{m_e c^2 r}\frac{\partial V}{\partial r}\mathbf{l} \qquad (6.50)$$

where $\mathbf{l}$ is the angular momentum of the particle whose mass is m_e that is, $\mathbf{l} = m_e\mathbf{r} \times \mathbf{v}$. If the magnetic moment of the moving particle is μ_s the energy of interaction (W) between this and the magnetic field is given by

$$W = -\tfrac{1}{2}\mu_s \cdot \mathbf{B} \qquad (6.51)$$

where the factor of one-half arises from a relativistic effect known as Thomas precession, which we shall not discuss further here except to note that it has nothing to do with the other relativistic factor of one-half mentioned in connection with Eq. (6.47). Combining (6.50), (6.51), and (6.47) we get

$$W = -\frac{e}{2m_e^2 c^2}\frac{1}{r}\frac{\partial V}{\partial r}\mathbf{l}\cdot\mathbf{s} \tag{6.52}$$

If the total angular momentum is **j**, then

$$j^2 = l^2 + s^2 + 2\mathbf{l}\cdot\mathbf{s} \tag{6.53}$$

and hence, combining (6.52) and (6.53)

$$W = -\frac{e}{4m_e^2 c^2}\frac{1}{r}\frac{\partial V}{\partial r}(j^2 - l^2 - s^2) \tag{6.54}$$

We can make the transformation from classical mechanics to quantum mechanics in the usual way by replacing the dynamical variables with appropriate operators, thus obtaining the following expression for the operator $\hat{H}$ representing the total energy of a one-electron atom (cf. (5.22))

$$\hat{H} = \hat{H}_v + \hat{l}^2/2mr^2 + f(r)(\hat{J}^2 - \hat{L}^2 - \hat{S}^2) \tag{6.55}$$

where

$$f(r) = -\frac{e}{4m_e^2 c^2 r}\frac{\partial V}{\partial r} \tag{6.56}$$

and $\hat{H}_0$ depends only on the radial coordinate r.

The operators $\hat{J}^2$, $\hat{L}^2$, and $\hat{S}^2$ all commute with each other, and none of them depends on r, so it follows that $\hat{H}$ commutes with all three angular-momentum operators and the angular and spin-dependent parts of the energy eigenfunctions are therefore eigenfunctions of $\hat{J}^2$, $\hat{L}^2$, and $\hat{S}^2$. But it is just such eigenfunctions that we obtained in the last section as linear combinations of simple products of orbital and spin functions. Following standard notation for the spin-orbit problem, we now denote the quantum numbers associated with the magnitude of the orbital and spin angular momenta by l and s respectively (corresponding to l_1 and l_2 above) and those associated with the magnitude and z component of the total angular momentum by j and m_j (instead of l and m). Values of j range in integer steps from $|l - s|$ to $l + s$. It is clear from the form of (6.53) that the separation of variables method can be used and the energy eigenfunctions u_{njlsm_j} are therefore written as products of a radial part, R_{njls}, and an angular and spin dependent part ϕ_{jlsm}, which is a linear combination of products of the appropriate spherical harmonics and spin vectors. The variables can then be separated directly and the energy levels determined from the resulting radial equation

$$\hat{H}_r R_{njls}(r) = E_{njls}R_{njls}(r) \tag{6.57}$$

where

$$\hat{H}_r = \hat{H}_0 + l(l + 1)\hbar^2/2mr^2 + \hat{H}'$$

and

$$\hat{H}' = f(r)[j(j+1) - l(l+1) - s(s+1)]\hbar^2 \qquad (6.58)$$

The inclusion of the spin-orbit coupling term therefore results in states having the same value of l, but different j, and which would otherwise be degenerate, having different energies.

The energy levels E_{njls} in (6.57) can be expressed as

$$E_{njls} = E_{nl} + \delta E_{njls} \qquad (6.59)$$

where the E_{nl} are what the energies would be in the absence of the spin-orbit interaction—i.e. they are the eigenvalues of the operator composed of the sum of the first two terms of (6.58). In the next chapter we shall describe a method for calculating such small corrections to energy eigenvalues when the uncorrected energy and eigenfunctions are known. We shall find that an energy correction such as δE_{njls} can be approximated by the expectation value of the additional part of the Hamiltonian—$\hat{H}'$ in (6.58) — with respect to the eigenfunctions of the 'unperturbed' Hamiltonian $(\hat{H}_r - \hat{H}')$. Applying these techniques to the present problem gives

$$\delta E_{njls} = \langle f(r)\rangle[j(j+1) - l(l+1) - s(s+1)]\hbar^2 \qquad (6.60)$$

where the expectation value $\langle f(r)\rangle$ is calculated using the radial part of the uncorrected wave function; in the case of hydrogen, for example, this is just R_{nl} as defined in (3.69) (see Problems 7.7 and 7.8).

In the case of a one-electron atom, s equals $\frac{1}{2}$ so the possible values of j are $l + \frac{1}{2}$ and $l - \frac{1}{2}$, unless l equals zero when j and s are identical. Thus it follows from (6.58) that the spin-orbit coupling results in all states with $l \neq 0$ being split into two, one member of the resulting doublet having its energy raised by an amount proportional to l while the other is lowered by an amount proportional to $(l + 1)$ (see Fig. 6.2). States with $l = 0$, on the other hand, have $j = s = \frac{1}{2}$ and are not split by spin-orbit coupling. We note that the quantum number m_j does not enter (6.58)—as indeed would be expected because the spherical symmetry of the problem has not been broken—so each state with a given value of j consists of $(2j + 1)$ degenerate components which can be separated by an applied magnetic field as discussed in the next section.

A familiar example of such a doublet is observed in the spectrum of sodium and other alkali metals. The outer electron in sodium can be considered as moving in a spherically symmetric potential resulting from the nucleus and 10 inner electrons. The well-known D lines result from transitions between states with $l = 1$ and $l = 0$ respectively. The first of these is split into two by the spin-orbit coupling, one component having $j = \frac{3}{2}$ and being fourfold degenerate and the other having $j = \frac{1}{2}$ and being doubly degenerate, while the second is not affected by spin-orbit coupling and is therefore single. Transitions between the two sets of levels therefore produce a pair of closely

Figure 6.2 The effect of spin-orbit coupling on the energy levels of a one-electron atom.

spaced spectral lines in agreement with experimental observation. The difference between the frequencies of these lines can be calculated in the way described above and such calculations produce results that are in excellent agreement with experiment in every case.

Similar effects of spin-orbit coupling are observed in the spectrum of hydrogen and it should be noted that these must be allowed for when a value of the Rydberg constant (cf. Chapter 3) is derived from measurements on these spectral lines.

6.6 THE ZEEMAN EFFECT

We now return to the problem referred to in Chapter 5 when we considered the effect of applying a magnetic field to an atom and found that this led to the degenerate energy levels being split in a way that could not be explained using orbital angular momentum alone. We shall now discuss this problem, paying particular attention to two extreme cases: the strong-field Zeeman effect where the energy of interaction between the atom and the applied field is much greater than the spin-orbit coupling and the weak-field Zeeman effect where the relative strengths are reversed.

If a magnetic field of magnitude B and directed along the z axis, is applied to a one-electron atom, further terms are introduced into the Hamiltonian which correspond to the interaction between this field and the magnetic moments associated with the orbital and spin angular momenta.

Expressions for these have been derived previously, and using (6.46), (6.47), (6.51), and (6.55) we obtain the following expression for the full Hamiltonian which includes both the Zeeman effect and spin-orbit coupling:

$$\hat{H} = \hat{H}_0 + \hat{L}^2/2mr^2 + f(r)(\hat{J}^2 - \hat{L}^2 - \hat{S}^2) + \frac{eB}{2m_e}(\hat{L}_z + 2\hat{S}_z) \quad (6.61)$$

In the case of the strong-field Zeeman effect, the last term is assumed to be much greater than the third and, as an initial approximation, the latter can be ignored. The energy eigenfunctions are therefore eigenfunctions of $\hat{L}^2$, $\hat{S}^2$, $\hat{L}_z$, and $\hat{S}_z$, but not $\hat{J}^2$, and so are just the simple product functions referred to in the discussion preceding Eq. (6.29). The energy eigenvalues are therefore changed by an amount equal to ΔE where

$$\Delta E = \frac{e\hbar B}{2m_e}(m_l + 2m_s) \quad (6.62)$$

and m_l and m_s are the quantum numbers associated with the z components of the orbital and spin angular momenta. Each state corresponding to given values of l and s is therefore split into $(2l + 1)(2s + 1)$ components, although some of these may be accidentally degenerate.

Although for typical field strengths the spin-orbit interaction is smaller than that representing the interaction with the field, it is rarely so small that it can be ignored completely and a further correction, δE, to each state specified by m_l and m_s is usually required. This can be calculated using the techniques of perturbation theory to be discussed in the next chapter in a manner similar to that referred to in the discussion of the spin-orbit interaction (cf. (6.60)). We get

$$\delta E = 2\langle f(r)\rangle m_l m_s \hbar^2 \quad (6.63)$$

where $\langle f(r)\rangle$ is the same quantity that appears in (6.60) (see Problem 7.9).

As an example we refer again to the state with $l = 1$ in an atom such as hydrogen or sodium with a strong magnetic field applied in the z direction. As $l = 1$ the possible values of m_l and m_s are ± 1 or 0, and $\pm\frac{1}{2}$ respectively so that $(m_l + m_s)$ can equal ± 2, ± 1 or 0 where the last state is doubly degenerate. We note that it follows from (6.63) and (6.60) that the ratio of the spin-orbit correction δE to the Zeeman-split-levels to the spin-orbit splitting $(\delta E_{n\,3/2\,1\,1/2} - \delta E_{n\,1/2\,1\,1/2})$ in zero field is equal to $2m_l m_s/3$. The resulting energy-level diagram in the zero-field and strong-field cases (along with the weak-field case to be discussed in the next paragraph) is illustrated in Fig. 6.3.

Turning now to the weak-field Zeeman effect, we saw above that when the last term in (6.61) is zero, the energy eigenfunctions are also eigenfunctions of $\hat{J}^2$, $\hat{L}^2$, $\hat{S}^2$ and $\hat{J}_z$. For very weak fields we can assume the eigenfunctions to be effectively unchanged† and each level with a particular

---

†This statement follows directly from perturbation theory as discussed in the next chapter.

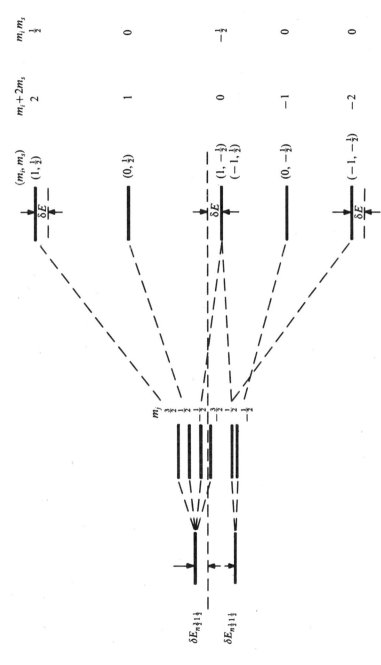

Figure 6.3 The energies of the states with $l = 1$ in a one-electron atom with a zero (left), weak (centre) and strong (right) magnetic field applied.

value of j, l and s is therefore split into $(2j + 1)$ states, each denoted by a particular value of m_j. This is illustrated in Fig. 6.3 which also indicates how the levels 'cross over' as the magnetic field is increased from zero to the strong-field limit.

The evaluation of the magnitude of the splitting in the weak-field case is complicated by the fact that the operator representing the total magnetic moment $(e/2m_e)(\hat{L} + 2\hat{S})$ represents a vector which is not in the same direction as the operator representing the total angular momentum $-(\hat{L} + \hat{S})$. A rigorous calculation involves manipulation of Clebsch–Gordan coefficients, but the same result can be obtained by the following physical argument. If we assume the field to be so weak that the eigenfunctions are unchanged from the zero-field case, components of the magnetic moment in directions other than that of the total angular momentum will have no effect on the interaction energy, which therefore results only from the component μ_j of the total magnetic moment in the direction of the total angular momentum. We first refer to the classical case where

$$\mu_j = \mathbf{j} \cdot \boldsymbol{\mu}/j$$

$$= \frac{e}{2m_e j} \mathbf{j} \cdot (\mathbf{l} + 2\mathbf{s}) = \frac{e}{2m_e j} (j^2 + \mathbf{j} \cdot \mathbf{s})$$

$$= \frac{ej}{2m_e} \left(1 + \frac{\mathbf{l} \cdot \mathbf{s} + s^2}{j^2}\right) = \frac{ej}{2m_e} \left(1 + \frac{j^2 - l^2 + s^2}{2j^2}\right)$$

using (6.53). The interaction energy δE is then given by $B\mu_{jz}$ where μ_{jz} is the z component of μ_j. That is

$$\delta E = \frac{eBj_z}{2m_e} \left(1 + \frac{j^2 - l^2 + s^2}{2j^2}\right) \tag{6.64}$$

We now replace the classical expressions for the magnitudes of the angular momenta by the eigenvalues of the corresponding quantum mechanical operators, so that the change in energy relative to the zero-field case is given by

$$\delta E = g\mu_B m_j B \tag{6.65}$$

where $\mu_B = e\hbar/2m_e$ is the Bohr magneton (5.27) and

$$g = 1 + \frac{j(j+1) - l(l+1) + s(s+1)}{2j(j+1)} \tag{6.66}$$

is known as the Landé g-factor. We see, for example, that in the case where $s = 0$, j must equal l and therefore g equals 1. This corresponds to all the angular momentum being orbital so that the total magnetic moment is directly proportional to the total angular momentum. This also happens if there is no orbital angular momentum (that is, $l = 0$ and therefore $j = s$) and in this case $g = 2$.

Although we have considered only a few comparatively simple examples, the principles underlying the addition of angular momenta have wide application. For example, such techniques can be applied to the problem of the coupling of the orbital and spin angular momenta of many-electron atoms when detailed calculations show that it can often be considered as a two-stage process: firstly the orbital and spin angular momenta of the different electrons separately combine to form total orbital and spin vectors, and these two quantities then interact in a similar way to that described above in the case of one-electron atoms. Such a system is described as Russell–Saunders or *L-S* coupling, and provides a satisfactory account of the atomic spectra of atoms with low atomic number. Heavy elements, on the other hand, are better described by '*j-j* coupling' in which the orbital and spin angular momenta of each electron interact strongly to form a total described by the quantum number j. The totals associated with different electrons then combine together to form a grand total for the atom.

The applications of many of the ideas discussed in this chapter and the last to the fields of nuclear and particle physics are also of great importance. This is not only due to the importance of angular momentum in these systems, but also because other physical quantities turn out to be capable of representation by a set of operators obeying the same algebra as do those representing angular momentum components. An example of this is the quantity known as *isotopic spin*. Using this concept the proton and neutron can be described as separate states of the same particle, assuming that the total isotopic-spin quantum number is $\frac{1}{2}$ so that its 'z component' can be represented by a quantum number having the values $\frac{1}{2}$ and $-\frac{1}{2}$. Properties such as the charge and mass differences between the two states can then be treated as resulting from isotopic-spin-dependent interactions between the three quarks that constitute the nucleon. It is also found that excited states of the nucleon exist whose properties can be explained by assigning appropriate values to the two isotopic-spin quantum numbers.

PROBLEMS

6.1 Show that the Hermitian matrices

$$[x] = \frac{1}{2}\left(\frac{\hbar}{m\omega}\right)^{1/2} \begin{bmatrix} 0 & 1 & 0 & 0 & \cdots \\ 1 & 0 & \sqrt{2} & 0 & \cdots \\ 0 & \sqrt{2} & 0 & \sqrt{3} & \cdots \\ 0 & 0 & \sqrt{3} & 0 & \cdots \\ \cdot & \cdot & \cdot & \cdot & \cdots \\ \cdot & \cdot & \cdot & \cdot & \cdots \end{bmatrix}$$

and

$$[P] = \tfrac{1}{2}(m\omega\hbar)^{1/2} \begin{bmatrix} 0 & -i & 0 & 0 & \cdots \\ i & 0 & -i\sqrt{2} & 0 & \cdots \\ 0 & i\sqrt{2} & 0 & -i\sqrt{3} & \cdots \\ 0 & 0 & i\sqrt{3} & 0 & \cdots \\ \cdot & \cdot & \cdot & \cdot & \cdots \\ \cdot & \cdot & \cdot & \cdot & \cdots \end{bmatrix}$$

obey the correct commutation rules for position and momentum in one dimension. Show that these can be used to obtain the energy levels of a harmonic oscillator of classical frequency ω and compare your answers with those obtained in Chapter 2.

6.2 Calculate the expectation values $\langle \hat{S}_x \rangle$, $\langle \hat{S}_y \rangle$, $\langle \hat{S}_x^2 \rangle$, and $\langle \hat{S}_y^2 \rangle$ for a spin-half particle known to be in an eigenstate of $\hat{S}_z$. Show that the product $\langle \hat{S}_x^2 \rangle \langle \hat{S}_y^2 \rangle$ is consistent with the uncertainty principle.

6.3 Obtain matrices representing the raising and lowering operators in the case of a spin-half system. Verify that these have the general properties attributed to ladder operators in Chapter 5.

6.4 A spin-half particle, initially in an eigenstate of $\hat{S}_x$ with eigenvalue $\tfrac{1}{2}\hbar$, enters a Stern–Gerlach apparatus oriented to measure its angular momentum component in a direction whose orientation with respect to the x and z axes is defined by the spherical polar angles θ and ϕ. Obtain expressions for the probabilities of obtaining the results $+\tfrac{1}{2}\hbar$ and $-\tfrac{1}{2}\hbar$ from the second measurement.

6.5 Spin-half particles, initially in an eigenstate of $\hat{S}_z$ with eigenvalue $\tfrac{1}{2}\hbar$, are directed into a Stern–Gerlach apparatus oriented to measure the x component of spin. The lengths of the two possible paths through the second apparatus are precisely equal and the particles are then directed into a common path so that it is impossible to tell which route any particular particle followed. What result would you expect to obtain if the z component of the particle spin is now measured? Discuss this from the point of view of the quantum theory of measurement and compare your discussion with that of the two-slit experiment discussed in Chapter 1.

6.6 Show that the following matrices obey the appropriate commutation rules and have the correct eigenvalues to represent the three components of angular momentum of a spin-one particle:

$$[L_x] = \frac{\hbar}{\sqrt{2}} \begin{bmatrix} 0 & 1 & 0 \\ 1 & 0 & 1 \\ 0 & 1 & 0 \end{bmatrix} \quad [L_y] = \frac{\hbar}{\sqrt{2}} \begin{bmatrix} 0 & -i & 0 \\ i & 0 & -i \\ 0 & i & 0 \end{bmatrix} \quad [L_z] = \hbar \begin{bmatrix} 1 & 0 & 0 \\ 0 & 0 & 0 \\ 0 & 0 & -1 \end{bmatrix}$$

Verify that the corresponding matrix representing the square of the total angular momentum also has the correct eigenvalues.

6.7 Obtain the eigenvectors of the matrices given in Prob. 6.6 and use these to find the relative probabilities of the possible results of the measurement of the x component of spin on a spin-one particle initially in an eigenstate of S_z with eigenvalue $\hbar$.

6.8 Draw up tables similar to Table 6.2 in the cases where (i) $l_1 = 3, l_2 = 1$, and (ii) $l_1 = 1, l_2 = \tfrac{3}{2}$.

6.9 Discuss the splitting of the $l = 2$ state of a one-electron atom due to (i) spin-orbit coupling, (ii) the strong-field Zeeman effect, and (iii) the weak-field Zeeman effect.

6.10 Discuss spin-orbit coupling and the strong- and weak-field Zeeman effects in the case of an energy level of helium whose total orbital and total spin quantum numbers are 2 and 1 respectively.

Hint: use Table 6.2.

SEVEN

TIME-INDEPENDENT PERTURBATION THEORY AND THE VARIATIONAL PRINCIPLE

Throughout our discussion of the general principles of quantum mechanics we have emphasized the importance of the eigenvalues and eigenfunctions of the operators representing physical quantities. However, we have seen that the solutions of the basic eigenvalue equations determining these are often not straightforward. In the early chapters discussing the energy eigenvalue equation (or time-independent Schrödinger equation) for example, we found that this could often not be solved exactly and that, even when a solution was possible, it frequently required considerable mathematical analysis. Because of these problems, methods have been developed to obtain approximate solutions to eigenvalue equations; one of the most important of such techniques is known as time-independent perturbation theory and will be the first method to be discussed in the present chapter, while another, known as the variational principle, will be discussed later. We shall confine our discussion to the particular case of the energy eigenvalue equation and use the wave function representation, but the results are readily generalized to other eigenvalue equations, expressed in either differential operator or matrix form.

Perturbation theory can be applied when the Hamiltonian operator $\hat{H}$, representing the total energy of the system, can be written in the form

$$\hat{H} = \hat{H}_0 + \hat{H}' \qquad (7.1)$$

where the eigenvalues E_{0n} and eigenfunctions u_{0n} of $\hat{H}_0$ are assumed to be known and the operator $\hat{H}'$ represents an additional energy known as a *perturbation*. Thus if we know the solution to a problem described by $\hat{H}_0$

(for example the energy eigenvalues and eigenfunctions of the hydrogen atom) we can use perturbation theory to obtain approximate solutions to a related problem (such as the energy eigenvalues and eigenfunctions of a hydrogen atom subject to a weak electric field). We shall use (7.1) along with the eigenvalue equation for the unperturbed system

$$\hat{H}_0 u_{0n} = E_{0n} u_{0n} \qquad (7.2)$$

to express the eigenvalues and eigenfunctions of $\hat{H}$ in the form of a series, the leading (zero-order) term of which is independent of $\hat{H}'$ while the next (first-order) term contains expressions linear in $\hat{H}'$ and so on; a similar series is also obtained for the energy eigenfunctions. The mathematical complexity of the series increases rapidly with ascending order, and perturbation theory is generally useful only if the series converges rapidly—which usually means that the additional energy due to the perturbation is much smaller than the energy difference between typical neighbouring levels. Accordingly, we shall confine our treatment to obtaining expressions for the eigenvalues of $\hat{H}$ which are correct to second order in $\hat{H}'$ as well as first-order corrections to the eigenfunctions. We shall illustrate the perturbation method by considering its application to several physical problems.

7.1 PERTURBATION THEORY FOR NON-DEGENERATE ENERGY LEVELS

We first consider the effect of perturbations on energy levels that are not degenerate and return to the degenerate case later. We assume that the correct eigenvalues and eigenfunctions of $\hat{H}$ can be expressed as a series whose terms are of zeroth, first, second, etc., order in the perturbation $\hat{H}'$, and derive expressions for each of these in turn. This can be done most conveniently if we rewrite (7.1) in the form

$$\hat{H} = \hat{H}_0 + \beta\hat{H}' \qquad (7.3)$$

where β is a constant, and obtain results that are valid for all values of β, including the original case where $\beta = 1$. The series for E_n and u_n can then be written as

$$\left. \begin{aligned} E_n &= E_{0n} + \beta E_{1n} + \beta^2 E_{2n} + \cdots \\ u_n &= u_{0n} + \beta u_{1n} + \beta^2 u_{2n} + \cdots \end{aligned} \right\} \qquad (7.4)$$

where the terms independent of β are known as zeroth-order terms, those in β are first-order, those in β^2 second-order, and so on. We substitute these expressions into the energy eigenvalue equation

$$\hat{H} u_n = E_n u_n \qquad (7.5)$$

and obtain

$$(\hat{H}_0 + \beta\hat{H}')(u_{0n} + \beta u_{1n} + \beta^2 u_{2n} + \cdots)$$

$$= (E_{0n} + \beta E_{1n} + \beta^2 E_{2n} + \cdots)(u_{0n} + \beta u_{1n} + \beta^2 u_{2n} + \cdots) \qquad (7.6)$$

Expanding this equation and equating the coefficients of the different powers of β we get

$$\hat{H}_0 u_{0n} = E_{0n} u_{0n} \qquad (7.7)$$

$$\hat{H}' u_{0n} + \hat{H}_0 u_{1n} = E_{0n} u_{1n} + E_{1n} u_{0n} \qquad (7.8)$$

$$\hat{H}' u_{1n} + \hat{H}_0 u_{2n} = E_{0n} u_{2n} + E_{1n} u_{1n} + E_{2n} u_{0n} \qquad (7.9)$$

Equations (7.7), (7.8), and (7.9) are valid for all values of β, including the case $\beta = 1$ when the first- and second-order corrections to the energy levels and eigenfunctions are E_{1n}, E_{2n}, u_{1n}, and u_{2n} respectively. We note first that (7.7) is identical to (7.2) which is to be expected as the former refers to a perturbation of zero order and the latter describes the unperturbed system. We can obtain expressions for the first-order corrections from (7.8) by expressing u_{1n} as a linear combination of the complete set of unperturbed eigenfunctions u_{0k}:

$$u_{1n} = \sum_k a_{nk} u_{0k} \qquad (7.10)$$

Substituting (7.10) into (7.8) gives, after rearranging and using (7.7),

$$(\hat{H}' - E_{1n})u_{0n} = \sum_k a_{nk}(E_{0n} - E_{0k})u_{0k} \qquad (7.11)$$

We now multiply both sides of (7.11) by u_{0n}^* and integrate over all space, using the fact that the u_{0k} are orthonormal to get

$$E_{1n} = H'_{nn} \qquad (7.12)$$

where

$$H'_{nn} = \int u_{0n}^* \hat{H}' u_{0n}\, d\tau \qquad (7.13)$$

We have therefore obtained an expression for the first-order correction to the energy eigenvalues of the perturbed system in terms of the perturbation operator H' and the eigenfunctions of the unperturbed system, which are assumed to be known. We note that this first-order correction to the energy is just the expectation value of the perturbation operator calculated using the unperturbed eigenfunctions. We can now obtain an expression for the first-order correction to the eigenfunction by multiplying both sides of (7.11) by u_{0m}^* (where $m \neq n$) and again integrating over all space when we get (again using orthonormality)

$$a_{nm} = \frac{H'_{mn}}{E_{0n} - E_{0m}} \qquad m \neq n \qquad (7.14)$$

where H'_{mm} is defined by a similar equation to (7.13) with u^*_{0n} replaced by u^*_{0m} and is therefore a *matrix element* (cf. Chapter 6). Hence, using (7.4) and (7.10),

$$u_n = (1 + a_{nn})u_{0n} + \sum_{k \neq n} \frac{H'_{kn}}{E_{0n} - E_{0k}} u_{0k} + \text{higher order terms} \quad (7.15)$$

We shall now show that a_{nn} can be put equal to zero. To do this we note that (7.15) must be normalized. That is

$$1 = \int u^*_n u_n \, d\tau$$

$$= \int \left[(1 + a^*_{nn})u^*_{0n} + \sum_{k \neq n} a^*_{nk}u^*_{0k} \right] \left[(1 + a_{nn})u_{0n} + \sum_{k \neq n} a_{nk}u_{0k} \right] d\tau$$

Remembering again that the u_{0k} are orthonormal and retaining only first-order terms we get

$$1 + a^*_{nn} + a_{nn} = 1$$

Hence

$$a_{nn} = -a^*_{nn} \quad (7.16)$$

and a_{nn} is therefore an imaginary number which we can write as $i\gamma_n$ where γ_n is real. Hence

$$1 + a_{nn} = 1 + i\gamma_n \simeq \exp(i\gamma_n)$$

where the final equality is correct to first order as is (7.15). It follows that the factor $(1 + a_{nn})$ in the first term of (7.15) is equal to $\exp(i\gamma_n)$ to first order and the effect of this is simply to multiply u_{0n} by a phase factor. Referring to (7.13) we see that the first-order change in the energy eigenvalue is independent of the phase of u_{0n} and from (7.15) we see that the value of γ_n affects only the overall phase of the eigenfunction—at least as far as zero- and first-order terms are concerned. But we know that the absolute phase of an eigenfunction is arbitrary, so there is no loss of generality involved if we put $\gamma_n = 0$, leading to the following expression for the eigenfunction which is correct to first order:

$$u_n = u_{0n} + \sum_{k \neq n} \frac{H'_{kn}}{E_{0n} - E_{0k}} u_{0k} \quad (7.17)$$

Proceeding now to consider second-order terms, we first expand u_{2n} in terms of the unperturbed eigenfunctions

$$u_{2n} = \sum_k b_{nk}u_{0k} \quad (7.18)$$

and then substitute from (7.7), (7.10), and (7.18) into (7.9) which gives, after some rearrangement,

$$\sum_k b_{nk}(E_{0k} - E_{0n})u_{0k} + \sum_{k \neq n} a_{nk}(\hat{H}' - E_{1n})u_{0k} = E_{2n}u_{0n} \qquad (7.19)$$

Multiplying (7.19) by u_{0n}^* and integrating over all space leads to

$$E_{2n} = \sum_{k \neq n} a_{nk}H'_{nk}$$

$$= \sum_{k \neq n} \frac{H'_{kn}H'_{nk}}{E_{0n} - E_{0k}}$$

$$= \sum_{k \neq n} \frac{|H'_{kn}|^2}{E_{0n} - E_{0k}} \qquad (7.20)$$

where we have used (7.14) and the relation $H'_{kn} = H'^*_{nk}$, which follows from the definition of the matrix element and the fact that $\hat{H}'$ is a Hermitian operator. The above procedure can be continued to obtain expressions for the second-order change in the eigenfunction as well as higher-order corrections to both the eigenfunctions and the eigenvalues. The expressions rapidly become more complicated, however, and if the perturbation series does not converge rapidly enough for (7.12), (7.17), and (7.20) to be sufficient, it is usually better to look for some other method of solving the problem.

Example 7.1 The anharmonic oscillator We consider the case of a particle of mass m subject to a one-dimensional potential $V(x)$ where

$$V = \tfrac{1}{2}m\omega^2 x^2 + \gamma x^4 \qquad (7.21)$$

and we wish to calculate the energy of the ground state to first order in γ. If γ were zero, the potential would correspond to a harmonic oscillator of classical frequency ω whose energy levels were shown in Chapter 2 to be

$$E_{0n} = (n + \tfrac{1}{2})\hbar\omega \qquad (7.22)$$

The corresponding ground-state energy eigenfunction was also evaluated in Chapter 2 as

$$u_{00} = \left(\frac{m\omega}{\pi\hbar}\right)^{1/4} \exp\left(-\frac{m\omega}{2\hbar}x^2\right) \qquad (7.23)$$

The inclusion of the term γx^4 in the potential (7.21) changes the problem from a harmonic oscillator to an anharmonic oscillator. If γ is small, we can apply perturbation theory and the first-order correction to the ground-state energy is obtained by substituting from (7.23) into (7.12)

Table 7.1 The ground-state energy of an anharmonic oscillator calculated using first-order perturbation theory (E_1) and by numerical methods (E')

$g = \dfrac{2\hbar\gamma}{m^2\omega^3}$	$E_1/(\tfrac{1}{2}\hbar\omega)$	$E'/(\tfrac{1}{2}\hbar\omega)$
0.01	1.007 50	1.007 35
0.1	1.075	1.065
0.2	1.15	1.12

giving

$$E_{10} = \left(\frac{m\omega}{\pi\hbar}\right)^{1/2} \int_{-\infty}^{\infty} \gamma x^4 \exp\left(-\frac{m\omega}{\hbar}x^2\right) dx$$

$$= \frac{3\hbar^2}{4m^2\omega^2}\gamma \tag{7.24}$$

which, along with the zero-order term $\tfrac{1}{2}\hbar\omega$, constitutes the required expression for the ground-state energy. If we characterize the 'strength' of the perturbation by the dimensionless quantity g defined as $g = (2\hbar/m^2\omega^3)\gamma$, it follows from (7.22) and (7.24) that the ratio of the first-order correction to the unperturbed energy is just $3g/4$. The perturbed energies calculated in this way are shown in Table 7.1 for several values of g, along with values of the total energy (as a fraction of $\tfrac{1}{2}\hbar\omega$) calculated by solving the Schrödinger equation numerically with the potential (7.21). We see that when the total correction to the zero-order energy is around one per cent, the error involved in using first-order perturbation theory is less than 0.02 per cent of the total, and that the approximation is still correct to about 3 per cent when the total correction is about 10 per cent.

Example 7.2 The atomic polarizability of hydrogen The d.c. polarizability of an atom α is defined by the equation $\mu = \alpha\varepsilon_0 F$ where μ is the electric dipole moment induced by a steady uniform electric field F. We shall consider a hydrogen atom in its ground state for which the wave function is spherically symmetric so that the direction of the z axis can be chosen as parallel to the field. A field of magnitude F in this direction will contribute an extra term, $\hat{H}'$, to the Hamiltonian which can be treated as a perturbation. Thus from elementary electrostatics

$$\hat{H}' = eFz \tag{7.25}$$

where z is the coordinate of the electron with respect to the proton at the origin. Substituting from (7.25) into (7.17), the ground-state eigenfunction u_0 is given by

$$u_0 = u_{00} + eF \sum_{k \neq 0} \frac{z_{k0}}{E_{00} - E_{0k}} u_{0k} \qquad (7.26)$$

Classically, if the electron were at a position **r** with respect to the nucleus, the atom would have a dipole moment μ where $\mu = -e\mathbf{r}$. It follows that the quantum-mechanical operator representing the dipole moment has the same form and the expectation value of its z component is therefore $\langle \mu \rangle$ where

$$\langle \mu \rangle = -e \int u_0^* z u_0 \, d\tau$$

and the components in the x and y directions clearly vanish from symmetry considerations. Using (7.26) we have

$$\langle \mu \rangle = -e \int u_{00}^* z u_{00} \, d\tau + 2e^2 F \sum_{k \neq 0} \frac{|z_{k0}|^2}{E_{0k} - E_{00}} \qquad (7.27)$$

to first order in F, remembering that $z_{k0}^* = z_{0k}$. The first term on the right-hand side of (7.27) vanishes because u_{00} is spherically symmetric and z is an odd function, so it follows that the atomic polarizability, α, is given by

$$\alpha = \frac{2e^2}{\varepsilon_0} \sum_{k \neq 0} \frac{|z_{k0}|^2}{E_{0k} - E_{00}} \qquad (7.28)$$

An alternative derivation which yields an identical expression to (7.28) consists of considering the perturbation expression for the energy: the first-order contribution (7.13) is proportional to z_{00} and is therefore zero on symmetry grounds, and the second-order term (7.20) is proportional to F^2. The latter is then compared with the energy of an induced dipole in the applied field which equals $-\frac{1}{2}\alpha\varepsilon_0 F^2$.

The summation in (7.28) is over all the excited states of the hydrogen atom (including the continuum of unbound states) and considerable computational effort would be required to evaluate it exactly. However, we can calculate a maximum possible value or 'upper bound' for α using what is known as the Unsöld closure principle. In this case we replace each energy difference $E_{0k} - E_{00}$ in (7.28) by its smallest possible value, which is clearly the difference between the ground and first excited energy levels. Writing this quantity as ΔE we get

$$\varepsilon_0 \alpha \leqslant \frac{2e^2}{\Delta E} \sum_{k \neq 0} |z_{k0}|^2$$

$$= \frac{2e^2}{\Delta E} \left(\sum_k z_{0k} z_{k0} - |z_{00}|^2 \right) \qquad (7.29)$$

As mentioned above, the second term in (7.29) vanishes for symmetry reasons. To evaluate the first term we refer back to the discussion of matrix representations in Chapter 6, from which it follows that the quantities z_{0k} constitute the elements of a matrix representing the operator z. If we now apply the standard rules for multiplying matrices we see that the first term in (7.29) is equal to the leading diagonal element of a similar matrix representing the operator z^2. Hence

$$\sum_k z_{0k} z_{k0} = (z^2)_{00} = \int u_{00}^* z^2 u_{00} \, d\tau \tag{7.30}$$

$$= \frac{1}{\pi a_0^2} \int_0^{2\pi} \int_0^{\pi} \int_0^{\infty} \exp\left(-\frac{2r}{a_0}\right) r^4 \cos^2 \theta \sin \theta \, dr \, d\theta \, d\phi$$

$$= a_0^2 \tag{7.31}$$

where we have used spherical polar coordinates and the standard expression for the ground-state eigenfunction (cf. 3.70)).

The energy difference ΔE is obtained from the expressions (3.72) for the energy levels of the hydrogen atom

$$\Delta E = \frac{3e^2}{8(4\pi\varepsilon_0)a_0} \tag{7.32}$$

Substituting from (7.31) and (7.32) into (7.29) we get

$$\alpha \leqslant 64\pi a_0^3/3$$

$$= 67.02 a_0^3 \tag{7.33}$$

This maximum value is within 15 per cent of the experimental value of $57.8 a_0^3$. Because the upper bound can be evaluated using only the ground-state eigenfunction and the difference between the energies of the two lowest states, this approximation can be usefully applied to the calculation of polarizabilities of other systems where the full set of matrix elements and energy differences are not known and expressions such as (7.28) cannot therefore be evaluated. Moreover, as we shall see later in this chapter, the variational principle can often be used to generate a lower bound to α so that by combining both methods, a theoretical estimate can be made with a precision that is rigorously known.

7.2 PERTURBATION THEORY FOR DEGENERATE LEVELS

The application of perturbation theory to degenerate systems can often lead to powerful insights into the physics of the systems considered. This is because the perturbation often 'lifts' the degeneracy of the energy levels, leading to additional structure in the line spectrum. We saw examples of this in Chapter 6 where the inclusion of a spin-orbit coupling term split the previously degenerate energy levels (Fig. 6.2) and further splitting resulted from the application of magnetic fields (Fig. 6.3). We shall now develop a general method for extending perturbation theory to such degenerate systems.

The mathematical reason why we cannot apply the theory in the form developed so far to the degenerate case follows from the fact that if one or more of the energy levels E_{0k} in (7.15) is equal to E_{0n} then at least one of the denominators $(E_{0n} - E_{0k})$ in (7.15) would be zero, leading to an infinite value for the corresponding term in the series. To understand the reason why this infinity arises, we consider the case of two-fold degeneracy and take u_{01} and u_{02} to be two eigenfunctions of the unperturbed Hamiltonian $\hat{H}_0$ with eigenvalue E_{01}. It is important to remember (see the discussion at the end of Chapter 4) that any linear combination of u_{01} and u_{02} is also an eigenfunction of H_0 with the same eigenvalue so that if we had solved the unperturbed problem in a different way we could well have come up with a different pair of eigenfunctions. Now suppose that a small perturbation $\hat{H}'$ is applied and that as a result we have two states of slightly different energy whose eigenfunctions are v_1 and v_2; because the system is no longer degenerate these functions are completely defined and linear combinations of them are not energy eigenfunctions. Now imagine we remove the perturbation gradually. As it tends to zero, v_1 and v_2 will tend to eigenfunctions of the unperturbed system; let these be u_1' and u_2'. In general u_1' and u_2' will be different from u_1 and u_2, although the former quantities will of course be linear combinations of the latter. Hence, if we start from u_1 and u_2 the application of a very small perturbation has to produce large changes in the eigenfunctions to get us to v_1 and v_2. This is why the infinite terms appear in the perturbation series.

The way out of this apparent impasse is to use other methods to obtain the correct starting functions u_1' and u_2'. Referring back to (7.11) we see that, if u_{01} is one of a degenerate pair whose other member is u_{02}, then the coefficient a_{12} on the right-hand side is indeterminate (because $E_{01} = E_{02}$) and the infinity arises later in the derivation when we divide through by $E_{01} - E_{02}$ which equals zero. However, we have seen that the correct zero-order eigenfunction should in general be some linear combination of u_{01} and u_{02} which we can write as

$$v_0 = C_1 u_{01} + C_2 u_{02} \qquad (7.34)$$

where the constants C_1 and C_2 are to be determined. Using this in place of u_{01}, the equivalent equation to (7.11) becomes

$$(\hat{H}' - E_1)(C_1 u_{01} + C_2 u_{02}) = \sum_k a_{1k}(E_{01} - E_{0k})u_{0k} \qquad (7.35)$$

where the first-order correction to the energy of the degenerate states is now written as E_1. We now multiply both sides of (7.35) by u_{01}^* and integrate over all space, then repeat the process using u_{02}^* to obtain the following pair of equations:

$$\left.\begin{array}{l} (H'_{11} - E_1)C_1 + H'_{12}C_2 = 0 \\ H'_{21}C_1 + (H'_{22} - E_1)C_2 = 0 \end{array}\right\} \tag{7.36}$$

where we have assumed that u_{01} and u_{02} are orthogonal because it was shown in Chapter 4 that, although degenerate eigenfunctions need not be orthogonal, an orthogonal set can always be generated using Schmidt orthogonalization. Equations (7.36) can be solved if, and only if, the determinant of the coefficients is zero:

$$\begin{vmatrix} H'_{11} - E_1 & H'_{12} \\ H'_{21} & H'_{22} - E_1 \end{vmatrix} = 0 \tag{7.37}$$

The solution of (7.37) leads to two possible values for E_1 and, substituting these into (7.36), and using the condition that the zero-order eigenfunction must be normalized, leads to two sets of values for the coefficients C_1 and C_2. Thus we have found two zero-order functions that are linear combinations of u_{01} and u_{02} and which, in general, correspond to different first-order corrections to the energy, so that, as we expected, one effect of the perturbation may be to remove the degeneracy. These two linear combinations therefore represent the appropriate choice of zero-order eigenfunctions for the problem and they can be used, along with the unperturbed eigenfunctions for the other states, to evaluate higher-order corrections in a similar way as in the non-degenerate case.

The above discussion has been confined to the case of twofold degeneracy, but the extension to the general (say M-fold) case is quite straightforward. The correct zero-order eigenfunctions, v_{0m} can be written as linear combinations of the original eigenfunctions u_{0k} according to

$$v_{0m} = \sum_{k=1}^{M} C_{mk}u_{0k} \tag{7.38}$$

and the first-order corrections to the energy are obtained from the determinantal equation

$$\begin{vmatrix} H'_{11} - E_1 & H'_{12} & \cdots & H'_{1M} \\ H'_{21} & H'_{22} - E_1 & \cdots & H'_{2M} \\ \vdots & \vdots & & \vdots \\ H'_{M1} & H'_{M2} & \cdots & H'_{MM} - E_1 \end{vmatrix} = 0 \tag{7.39}$$

The coefficients C_{mk} are obtained by substituting the resulting values of E_1 into the generalized form of (7.36) and applying the normalization condition. Once the appropriate set of unperturbed eigenfunctions has been found, higher-order corrections can be applied, just as in the non-degenerate case.

However, in most cases, the results of physical interest emerge from the first-order treatment just described.

Example 7.3 The Stark effect in hydrogen The effect of applying an electric field to the hydrogen atom in its ground state was discussed earlier in this chapter where we showed that there was no first-order change in the energy of this state. We now consider the first excited state which is fourfold degenerate† in the absence of any perturbation and we shall find that this level is split into three when an electric field is applied. This is known as the *Stark effect*.

The four degenerate states corresponding to the unperturbed first excited state of hydrogen all have quantum number n equal to 2 and the eigenfunctions, referred to spherical polar coordinates (r, θ, ϕ), are (cf. Chapter 3, Eq. (3.70)):

$$
\left.
\begin{aligned}
u_1 &\equiv u_{200} = (8\pi a_0^3)^{-1/2}(1 - r/2a_0)\, e^{-r/2a_0} \\
u_2 &\equiv u_{210} = (8\pi a_0^3)^{-1/2}(r/2a_0)\cos\theta\, e^{-r/2a_0} \\
u_3 &\equiv u_{211} = -(\pi a_0^3)^{-1/2}(r/8a_0)\sin\theta\, e^{i\phi}\, e^{-r/2a_0} \\
u_4 &\equiv u_{21-1} = (\pi a_0^3)^{-1/2}(r/8a_0)\sin\theta\, e^{-i\phi}\, e^{-r/2a_0}
\end{aligned}
\right\}
\tag{7.40}
$$

The perturbation due to the applied field (assumed as usual to be in the z direction) is represented by $\hat{H}'$ where (cf. (7.25))

$$
\hat{H}' = eFz = eFr\cos\theta
\tag{7.41}
$$

If we now proceed to evaluate the matrix elements we find that most of them vanish because of symmetry. Thus $H'_{11} = H'_{22} = H'_{33} = H'_{34} = H'_{43} = H'_{44} = 0$ because in each case the integrand is antisymmetric in z, and $H'_{13} = H'_{14} = H'_{23} = H'_{24} = H'_{31} = H'_{32} = H'_{41} = H'_{42} = 0$ because these integrands all contain the factor $e^{i\phi}$ and the integration with respect to ϕ is from $\phi = 0$ to $\phi = 2\pi$. This leaves only the matrix elements H'_{12} and H'_{21} which are equal to each other and given by

$$
\begin{aligned}
H'_{12} &= H'_{21} \\
&= \int_0^{2\pi}\int_0^{\pi}\int_0^{\infty} (8\pi a_0^3)^{-1}(r/2a_0)(1 - r/2a_0) \\
&\quad \times eFr\cos^2\theta\, e^{-r/a_0}r^2\, dr\,\sin\theta\, d\theta\, d\phi \\
&= \frac{eF}{8a_0^4}\int_0^{\pi}\cos^2\theta\sin\theta\, d\theta\int_0^{\infty}(r^4 - r^5/2a_0)\, e^{-r/a_0}\, dr \\
&= -3eFa_0
\end{aligned}
\tag{7.42}
$$

----

† For the purposes of this discussion we shall assume that the perturbation associated with the applied electric field is much larger than the splitting resulting from spin-orbit coupling discussed in the previous chapter so that the latter effect can be ignored.

The determinantal equation (7.39) therefore becomes in this case

$$
\begin{vmatrix}
-E_1 & -3eFa_0 & 0 & 0 \\
-3eFa_0 & -E_1 & 0 & 0 \\
0 & 0 & -E_1 & 0 \\
0 & 0 & 0 & -E_1
\end{vmatrix} = 0 \qquad (7.43)
$$

leading to the following expressions for the first-order corrections to the energy and for the zero-order eigenfunctions—making use of the generalized form of (7.36)—

$$
\left.
\begin{aligned}
E_1 &= 3eFa_0 & v_1 &= \frac{1}{\sqrt{2}}(u_1 - u_2) \\[2ex]
E_1 &= -3eFa_0 & v_2 &= \frac{1}{\sqrt{2}}(u_1 + u_2) \\[2ex]
E_1 &= 0 & v_3 &= u_3 \quad \text{and} \quad v_4 = u_4
\end{aligned}
\right\} \qquad (7.44)
$$

We should note several points about these results. First, the degeneracy of the last two states has not been lifted by the perturbation so any linear combination of u_3 and u_4 is a valid eigenfunction with $E_1 = 0$. Second, we see from Fig. 7.1, which shows a plot of the probability densities $|v_1^2|$ and $|v_2^2|$, that these are not symmetric across the plane

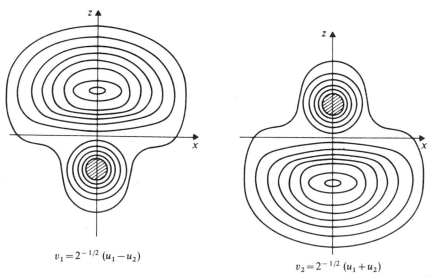

$$v_1 = 2^{-1/2}(u_1 - u_2)$$

$$v_2 = 2^{-1/2}(u_1 + u_2)$$

Figure 7.1 A section at $y = 0$ through the position probability distribution corresponding to the $n = 2$ energy state of a hydrogen atom subject to an electric field in the z direction. (A number of contours have been omitted in the high peaks represented by the cross-hatched areas.)

$z = 0$, implying that in these states the atom possesses a dipole moment that is aligned antiparallel to the field in the first case and parallel to it in the second. The energy changes can therefore be thought of as arising from the interaction between these dipoles and the applied field. However, in contrast to the polarization of the ground state discussed earlier (Sec. 7.1) these dipole moments are not generated by the physical operation of the field and their creation does not require the expenditure of any energy: the polarized states are simply two of the possible eigenfunctions of the degenerate unperturbed system. The energy change is therefore proportional to the field magnitude F, rather than to F^2 as is the case for the ground state. Third, we see from (7.44) that neither v_3 nor v_4 (and also by implication no linear combination of these functions) possesses a dipole moment and therefore neither has a first-order interaction with the field; the energy of these states is not affected by the perturbation to first order and they remain degenerate even in the presence of the field. Finally, we note that the occurrence of the Stark effect is a result of a mixing of the wave functions associated with the spherically symmetric $2s$ state and the $2p$ state with $m = 0$. In the hydrogenic atom these states are degenerate but this is not the case in other atomic systems where a linear Stark effect is therefore not generally found.

Experimental studies of the Stark effect have confirmed the theoretical predictions discussed above: the spectral lines of hydrogen are split by an electric field, the splitting being proportional to the field strength as expected, and the constants of proportionality agree with those calculated from (7.44) in the case of the $n = 2$ to $n = 1$ transition, and with the corresponding formulae in the case of other transitions. Moreover, the effect of an electric field on the spectra of most other atoms is to shift the spectral lines by an amount proportional to the square of the field strength; apart from hydrogen, a linear splitting is observed only in the case of a few atoms such as helium where the energy difference between the s and p states is small compared with the perturbation resulting from a strong electric field (cf. Chapter 10).

Example 7.4 Electrons in a one-dimensional solid We know from the fact that metals can carry electric currents that they contain electrons that are mobile and not attached to particular atoms; such electrons are known as 'free electrons'. We also know, on the other hand, that not all electrons are free and that many solids are insulators. In this example we shall develop a one-dimensional model of a solid in which we study the behaviour of otherwise free electrons in a potential similar to that arising from the nuclei and other electrons in a real solid. We shall assume that this potential is weak enough to be treated as a perturbation and we shall see that this simple model provides quite a good explanation of the electrical conductivity of solids.

The atoms in a crystalline solid are arranged on a regular lattice, so, in one dimension, we expect the potential experienced by a free electron to vary periodically with distance. If we call this repeat distance a, the simplest form of such a periodic potential is

$$V(x) = V_0 \cos(2\pi x/a) \tag{7.45}$$

V_0 is assumed to be small so that V can be treated as a perturbation on the case of a free electron whose unperturbed eigenfunctions therefore have the form

$$u_k = L^{-1/2} \exp(ikx) \tag{7.46}$$

where the factor $L^{-1/2}$ ensures that the eigenfunctions are normalized when integrated over the distance L, which is taken as the length of the piece of solid being considered. If this contains N atoms, it follows that $L = Na$; we assume that N is large enough for L to be of macroscopic dimensions and we shall discuss the nature of the boundaries of the solid later. The unperturbed energies E_{0k} are given by (cf. Sec. 2.4)

$$E_{0k} = \frac{\hbar^2 k^2}{2m} \tag{7.47}$$

and the states u_{0k} and u_{0-k} are therefore degenerate. Applying degenerate perturbation theory, the appropriate matrix elements linking these states are given by

$$H'_{kk} = H'_{-k-k} = L^{-1} \int_0^{Na} V_0 \cos(2\pi x/a)\, dx$$

$$= 0 \tag{7.48}$$

$$H'_{k-k} = H'^*_{-kk} = L^{-1} \int_0^{Na} e^{-2ikx} V_0 \cos(2\pi x/a)\, dx$$

$$= \frac{V_0}{2L} \int_0^{Na} \left[e^{i(-2k + 2\pi/a)x} + e^{i(-2k - 2\pi/a)x} \right] dx$$

$$= 0 \text{ unless } k = \pm\frac{\pi}{a}$$

$$= \frac{V_0}{2} \text{ if } k = \pm\frac{\pi}{a} \tag{7.49}$$

It follows that there is no first-order change in the energy unless $k = \pm\pi/a$, and when either of these conditions is satisfied, Eq. (7.39) becomes

$$\begin{vmatrix} -E_1 & \frac{1}{2}V_0 \\ \frac{1}{2}V_0 & -E_1 \end{vmatrix} = 0 \tag{7.50}$$

Thus $E_1 = \pm\frac{1}{2}V_0$ and the degeneracy of those states where $k = \pm\pi/a$ is lifted in first order.

Turning now to the second-order contributions to the energy, the general formula for which is (7.20), it can be easily shown that the matrix element connecting the states labelled by k and k' is only non-zero if $k - k' = \pm 2\pi/a$ when its value is $\frac{1}{2}V_0$. For states whose value of k is far from the points $\pm\pi/a$, therefore, the energy difference $(E_{0k} - E_{0k'})$ which appears in the denominator of (7.20) is large compared with $\frac{1}{2}V_0$, and the second-order term will be small enough to be neglected. Moreover, when k is equal to $\pm\pi/a$ we have the degenerate case considered above and, provided we use the correct zero-order eigenfunctions, the second-order terms are again small enough to be ignored. If we now consider states whose values of k are in the near vicinity of $\pm\pi/a$, the matrix element of the perturbation will be much larger than the difference between the unperturbed energies of the states k and k' and we now have an example of a 'nearly degenerate' system in which the perturbation series diverges, but the techniques discussed above cannot be applied because the degeneracy is not exact. However, we might expect that we could make a smooth connection between these extreme cases and so obtain the curve of E_k as a function of k shown in Fig. 7.2. This procedure can be justified by solving the problem using other techniques that are outside the scope of this book (see, for example, J. R. Hook and H. E. Hall *Solid State Physics*, Wiley, New York, 1991). A notable feature of this diagram is the appearance of gaps of width V_0 in the energy spectrum corresponding to the points $k = \pm\pi/a$. The influence of these gaps on the physical properties of the system is considerable and we shall discuss this shortly. First, however, we must consider more carefully the boundary conditions to be imposed on the problem.

The length L above was described as the length of the piece of solid under consideration so the most obvious boundary condition would be to require the wave function to be zero outside the range between $x = 0$ and $x = L$. Such 'fixed boundary conditions' can be used, but they give rise to considerable mathematical complications associated with the fact that plane waves of the form (7.46) are no longer eigenfunctions of the unperturbed system. It is therefore more convenient to use 'periodic boundary conditions' where we impose the condition that $u_k(x) = u_k(x + L)$ and the allowed values of k are therefore those where $k = 2n\pi/L$, n being an integer. This boundary condition corresponds physically to considering the one-dimensional line to be bent round into a closed loop, when the

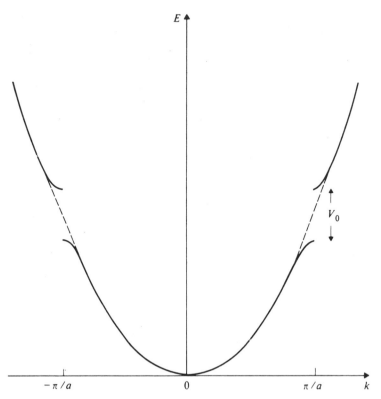

Figure 7.2 The energy of an electron in a one-dimensional metal as a function of wave number k, compared with that of a free electron (broken line), showing the energy gaps at $k = \pm\pi/a$.

allowed values of k follow from the condition that the wave function be single-valued. As bulk properties such as the electrical conductivity of real solids are not dependent on parameters such as the size and shape of the sample being considered, it is a reasonable assumption that the precise form of boundary conditions will not be important when discussing similar properties of our one-dimensional model. The most convenient form can therefore be chosen and this turns out to correspond to the periodic boundary conditions discussed above.

We saw above that the number of atoms associated with a length L of this one-dimensional solid is N where $L = Na$ and a is the repeat distance of the periodic potential; and that the number of energy states associated with values of k between plus and minus π/a is also N, because the periodic boundary conditions require that successive allowed values of k are separated by $2\pi/L$. It follows from the Pauli exclusion principle to be discussed in Chapter 10 that each state can be occupied by no more than two electrons, which must have opposite spin. In the case where each atom contributes one free electron to the system, the ground state of the system will therefore correspond to the ($N/2$) states of lowest energy being filled. If now an electric field is applied, some of the electrons will be excited so that there are more with (say) positive k than with negative k and, as k is proportional to the electron momentum, an electric current results. In this case the one-dimensional solid behaves like a metal. However, if there are two free electrons per atom, all the states in the band of energies whose k values lie between plus and minus π/a will be occupied. If V_0 is large enough, an applied field will be unable to excite any electrons, and there will therefore be as many electrons with positive k as there are with negative k; no current can then flow, and the one-dimensional solid in this case is an insulator. If, however, V_0 (and consequently the size of the energy gap) is small enough, some electrons will be thermally excited into states with $|k|$ greater than π/a where they will now be mobile and can respond to a field. This excitation leaves behind 'holes' in the otherwise full band and it can be shown that these have the same electrical properties as positively charged mobile particles. Such a system is a one-dimensional intrinsic semi-conductor.

The above discussion can be generalized to the case where there are more than two electrons per atom if we remember that a general periodic potential has Fourier components with repeat distances a, $a/2$, etc., giving rise to energy gaps at $k = \pm \pi/a$, $\pm 2\pi/a$, etc. It is then clear that any one-dimensional solid which possesses an odd number of free electrons per atom will be a metal, while an even number will imply insulating or semi-conducting properties. The arguments can be extended to three-dimensional solids where the rule is not quite so simple, but where quantum mechanics has been successfully used to explain the wide range of physical phenomena displayed by such materials. The interested reader should consult a textbook on solid state physics, such as that by Hook and Hall referred to above.

7.3 THE VARIATIONAL PRINCIPLE

This approximate method can be applied to eigenvalue problems where we know the operator $\hat{H}$ and can make some guess as to the form of the eigenfunction corresponding to the lowest energy state of the system. The variational principle can then be used to improve our guessed eigenfunction and produce a maximum value, or upper bound, to the ground-state energy. The usefulness of the variational principle depends strongly on our ability to make a good initial guess but, as we shall see, the symmetry and other physical properties of the system can often be useful guides to this. The variational principle can be extended to consider states other than the ground state of the system, but this has limited applications and will not be discussed here. We shall now proceed to describe the method in detail.

Let v be a function that is to approximate to the ground-state eigenfunction of the Hamiltonian operator $\hat{H}$. We shall now show that the expectation of $\hat{H}$ calculated using v cannot be less than the ground-state eigenvalue E_0. Allowing for the fact that v may not be normalized, this expectation value $\langle \hat{H} \rangle$ is given by

$$\langle \hat{H} \rangle = \frac{\int v^* \hat{H} v \, d\tau}{\int v^* v \, d\tau} \tag{7.51}$$

Using completeness, v can be expressed as a linear combination of the true, but unknown, eigenfunctions of $\hat{H}$:

$$v = \sum_n a_n u_n \tag{7.52}$$

Using (7.52) the numerator of (7.51) can be expressed as

$$\int v^* \hat{H} v \, d\tau = \int \left(\sum_n a_n^* u_n^* \right) \hat{H} \left(\sum_m a_m u_m \right) d\tau$$

$$= \sum_{nm} a_n^* a_m \int u_n^* \hat{H} u_m \, d\tau$$

$$= \sum_{nm} a_n^* a_m E_m \delta_{nm}$$

$$= \sum_n |a_n|^2 E_n \tag{7.53}$$

using orthonormality. If E_0 is the ground-state eigenvalue it must be smaller than all the other E_n's, so we can write

$$\int v^* \hat{H} v \, d\tau \geqslant E_0 \sum_n |a_n|^2 \tag{7.54}$$

The denominator of (7.51) can be similarly shown to be given by

$$\int v^* v \, d\tau = \sum_n |a_n|^2 \tag{7.55}$$

and combining (7.51), (7.54), and (7.55) we get

$$\langle \hat{H} \rangle \geqslant E_0 \tag{7.56}$$

as required. Thus by guessing an approximate eigenfunction v we can always calculate an upper bound to the ground-state energy. Moreover, if v contains adjustable parameters, these can be varied until $\langle \hat{H} \rangle$ has its smallest possible value when v will represent the best possible approximation of this form.

Example 7.5 The harmonic oscillator In this example we suppose that we do not know the ground-state eigenfunction for a one-dimensional harmonic oscillator, but have guessed that it is similar to that for a particle in an infinite potential well. That is,

$$\begin{aligned} v &= a^{-1/2}\cos(\pi x/2a) & a \leqslant x \leqslant a \\ &= 0 & |x| > a \end{aligned} \Bigg\} \tag{7.57}$$

We wish to find the value of a for which the expectation value of the energy is a minimum. Using (7.51)

$$\langle \hat{H} \rangle = a^{-1} \int_{-a}^{a} \cos(\pi x/2a) \left(-\frac{\hbar^2}{2m}\frac{\partial^2}{\partial x^2} + \frac{1}{2}m\omega^2 x^2 \right) \cos(\pi x/2a)\,dx$$

where m is the particle mass and ω the classical frequency of the oscillator. The differentiation and integration are straightforward, if a little tedious, and we get

$$\langle \hat{H} \rangle = \frac{\hbar^2\pi^2}{8ma^2} + m\omega^2 a^2 \left(\frac{1}{6} - \frac{1}{\pi^2} \right) \tag{7.58}$$

To find an expression for a corresponding to the minimum value of $\langle \hat{H} \rangle$ we put $\partial\langle \hat{H} \rangle/\partial a = 0$ and get

$$-\frac{\hbar^2\pi^2}{4ma^3} + 2m\omega^2 a \left(\frac{1}{6} - \frac{1}{\pi^2} \right) = 0$$

so that

$$\begin{aligned} a &= \pi \left[\frac{3}{4(\pi^2 - 6)} \right]^{1/4} \left(\frac{\hbar}{m\omega} \right)^{1/2} \\ &= 2.08 \left(\frac{\hbar}{m\omega} \right)^{1/2} \end{aligned} \tag{7.59}$$

We can substitute from (7.59) into (7.58) to obtain the minimum value of $\langle \hat{H} \rangle$ which we write as $\langle \hat{H} \rangle_{\text{min}}$. Thus

$$\langle \hat{H} \rangle_{\text{min}} = \frac{1}{2} \left(\frac{\pi^2 - 6}{3} \right)^{1/2} \hbar\omega$$

$$= 0.568\hbar\omega \tag{7.60}$$

Thus we have been able to set an upper bound to the energy of the oscillator which is within 14 per cent of the true value of $\frac{1}{2}\hbar\omega$. Figure 7.3 compares the approximate wave function evaluated using the value of a given in (7.59) with the exact ground-state eigenfunction obtained in Chapter 2.

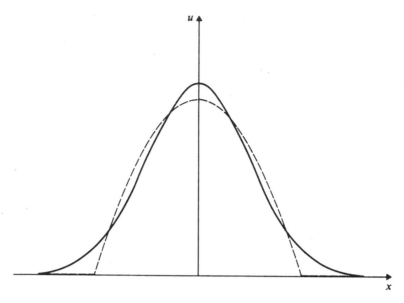

Figure 7.3 The wave function corresponding to the ground state of a one-dimensional harmonic oscillator (continuous line) compared with that for an infinite-sided well whose parameters are chosen using the variational principle (broken line).

Example 7.6 The atomic polarizability of hydrogen This problem was previously treated by perturbation theory where we showed (7.33) that an upper bound to the polarizability α is $64\pi a_0^3/3$. As the energy of an atom in an electric field of magnitude F is equal to $-\frac{1}{2}\alpha F^2$, this corresponds to a lower bound for the energy; we shall now use the variational principle to set an upper bound for the energy and hence a lower bound to α.

We showed earlier that the Hamiltonian for a hydrogen atom in a uniform electric field F in the positive z direction is

$$\hat{H} = \hat{H}_0 + eFz \tag{7.61}$$

We expect the application of an electric field to change the energy eigenfunction of an atom so as to enhance the probability of finding the electron at negative rather than positive values of z. A simple (unnormalized) function with these properties is

$$v = u_0(1 - \beta z) \tag{7.62}$$

where u_0 is the ground-state eigenfunction of $\hat{H}_0$ given in Chapter 3 (3.70) and β is a constant. We note that u_0 is a real function, and it turns out that we can also assume β to be real without any loss of generality. We can therefore write the expectation value of $\hat{H}$ as

$$\langle \hat{H} \rangle = \frac{\int u_0(1 - \beta z)(\hat{H}_0 + eFz)u_0(1 - \beta z)\, d\tau}{\int u_0^2(1 - \beta z)^2 \, d\tau} \tag{7.63}$$

Remembering that $\hat{H}_0 u_0 = E_0 u_0$, where E_0 is the ground-state energy in the zero-field case, and using the fact that $\hat{H}_0$ is Hermitian, we can show that all integrals that arise in the expansions of the numerator and denominator of (7.63), and which involve odd powers of z, must be equal to zero on symmetry grounds. Moreover, $\int u_0 z \hat{H}_0 z u_0 \, d\tau$ also vanishes, as can be shown by substituting the spherical polar expressions for the operators and eigenfunctions, while the integral $\int u_0 z^2 u_0 \, d\tau$ was earlier shown (7.31) to be equal to a_0^2. Equation (7.63) therefore becomes

$$\langle \hat{H} \rangle = \frac{E_0 - 2eF\beta a_0^2}{1 + \beta^2 a_0^2}$$

$$\simeq E_0(1 - \beta^2 a_0^2) - 2eF\beta a_0^2 \tag{7.64}$$

where we have ignored powers of β higher than the second as we are interested in the case of low fields where β is expected to be small. We can now differentiate to find a value of β corresponding to the minimum value of $\langle \hat{H} \rangle$:

$$-2a_0^2\beta E_0 - 2eFa_0^2 = 0$$

that is,

$$\beta = -\frac{eF}{E_0} \tag{7.65}$$

Substituting into (7.64), the minimum value $\langle \hat{H} \rangle_{\min}$ of $\langle \hat{H} \rangle$ is

$$\langle \hat{H} \rangle_{\min} = E_0 + \frac{e^2 F^2}{E_0} a_0^2$$

$$= E_0 - 8\pi\varepsilon_0 a_0^3 F^2 \tag{7.66}$$

using (3.72). This represents an upper bound to the ground-state energy, so the corresponding lower bound for the polarizability is $16\pi a_0^3$. We can combine this with the upper bound obtained by perturbation methods and the Unsöld approximation to give

$$16\pi a_0^3 \leqslant \alpha \leqslant 64\pi a_0^3/3 \tag{7.67}$$

Taking as a best estimate, the average of these two bounds, we get

$$\alpha = (18.8 \pm 2.8)\pi a_0^3 \tag{7.68}$$

and we have therefore obtained a theoretical value for the polarizability that is rigorously correct to within about 14 per cent. Moreover, it is in excellent agreement with experimental value of $18.4\pi a_0^3$.

A combination of perturbation and variational calculations can quite often be used in a similar way to that just described to obtain upper and lower bounds to the theoretical estimates of physical quantities that cannot be calculated directly—perhaps because the unperturbed eigenvalues and eigenfunctions are unknown. Such methods are frequently used, for example, in the calculation of the polarizabilities of non-hydrogen atoms and of molecules, and in the estimating of the 'van der Waals' interactions between atoms in gases. In favourable cases, by the use of sophisticated trial functions in the variational method and special techniques to estimate the sums over states in the perturbation expressions, the upper and lower bounds can be made to approach each other very closely so that highly accurate theoretical results can be obtained.

PROBLEMS

7.1 A particle moves in a potential given by

$$V = V_0 \cos(\pi x/2a) \quad (-a \leqslant x \leqslant a) \qquad V = \infty \quad (|x| > a)$$

where V_0 is small. Treat this problem as a perturbation on the case of a particle in an infinite-sided square well of length $2a$ and calculate the changes in the energies of the three lowest energy states to first order in V_0.

7.2 A particle moves in a potential given by

$$V = V_0 \quad (-b \leqslant x \leqslant b) \qquad V = 0 \quad (b < |x| \leqslant a) \qquad V = \infty \quad |x| > a$$

Calculate the energies of the three lowest states to first order in V_0 using a similar procedure to that in Prob. 7.1.

7.3 The hydrogenic atom was treated in Chapter 3 on the assumption that the nucleus has a point charge, while in reality the nuclear charge is spread over a small volume. Show that if we were to assume that the nuclear charge was in the form of a thin shell of radius δ, where $\delta \ll a_0$, the ground state energy of a hydrogen atom would be increased by an amount equal to $Ze^2\delta^2/(6\pi\varepsilon_0 a_0^3)$. Calculate this quantity for hydrogen assuming $\delta = 10^{-15}$ m and express your result as a fraction of the total ground-state energy.

7.4 A particle of mass m is attached by a massless rigid rod of length a to a fixed point and can rotate about a fixed axis passing through this point. Its energy eigenvalues and corresponding eigenfunctions are then $E_n = \hbar^2 n^2/2ma^2$ and $u_n = (2\pi)^{-1/2} \exp(in\phi)$ respectively (cf. Prob. 5.2). If this system is now perturbed by a potential $V_0 \cos(2\phi)$, calculate the first-order changes in the three lowest energy levels, remembering to allow for degeneracy where appropriate. Show that the second-order change in the ground ($n = 0$) state energy is equal to $-ma^2V_0^2/4\hbar^2$.

7.5 Show that a first-order Stark effect is possible only if the unperturbed eigenfunctions do not have a definite parity. How then is it possible to observe the Stark effect in hydrogen where the potential is centrosymmetric?

7.6 Use perturbation theory to show that the first-order changes in energy levels due to spin-orbit coupling are given by

$$[j(j + 1) - l(l + 1) - s(s + 1)]\hbar^2 \langle f(r) \rangle$$

in the notation of Chapter 6 where the expectation value $\langle f(r) \rangle$ is calculated using only the radial part of the unperturbed eigenfunction. Use the fact that the splitting of the D lines in sodium is about 6×10^{-10} m and that their mean wavelength is 5.9×10^{-7} m to estimate $\langle f(r) \rangle$ in the case of the $l = 1$ state involved in the D lines of sodium.

7.7 Use the result obtained in Prob. 7.6 to show that spin-orbit coupling splits the $2p$ level in hydrogen into a doublet whose separation, expressed as a fraction of the mean unperturbed energy, is equal to $(e^2/8\pi\varepsilon_0\hbar c)^2$.

7.8 Use perturbation theory to obtain the spin-orbit correction to the strong-field Zeeman levels (6.63). *Hint.* First show that the perturbation can be written in the form $2f(r)\hat{\mathbf{L}}\cdot\hat{\mathbf{S}}$; remember that the unperturbed eigenfunctions are eigenfunctions of $\hat{L}_z$ and $\hat{S}_z$.

7.9 Use the variational principle and the trial function $\exp(-\alpha x^2)$ to obtain an upper limit for the ground-state energy of a one-dimensional harmonic oscillator. Compare your results with the exact expressions given in Chapter 2.

EIGHT

TIME DEPENDENCE

So far in our development of quantum mechanics, we have given very little consideration to problems involving the time dependence of the wave function. This is rather surprising as in classical mechanics it is time-dependent phenomena—i.e. dynamics rather than statics—which command most attention. Moreover, most experimental observations necessarily involve some change in the quantity being observed and we might therefore have expected that the successful prediction of experimental results would require a detailed understanding of the way a system changed in time. In fact we have made some implicit assumptions about time dependence and the principal reason why we have got so far without discussing it explicitly (apart from a brief introduction in Chapter 4) is that many observed quantum phenomena are associated with sudden discontinuous changes between otherwise stable states. Thus most of our information concerning the energy levels of atoms has been obtained from measurements of the frequencies of electromagnetic radiation emitted or absorbed as the atom undergoes a transition from one energy eigenstate to another, assuming the correctness of the formula $E = \hbar\omega$, but not considering the mechanism of the transition in any detail. Assumptions concerning time dependence are also implicit in the quantum theory of measurement which refers to the probability of obtaining a particular result *following* a measurement performed on a system in a given state; thus, for example, the state vector of a spin-half particle changes from being an eigenvector $\hat{S}_z$ to being one of $\hat{S}_x$ following a measurement of the latter property using an appropriately oriented Stern–Gerlach apparatus.

There are, however, a number of problems in which the time dependence of the wave function must be considered explicitly, and some of these will be discussed in the present chapter. We already know from Postulate 4.5 that the basic equation governing the time evolution of the wave function is the time-dependent Schrödinger equation

$$i\hbar \frac{\partial \Psi}{\partial t} = \hat{H}\Psi \tag{8.1}$$

where $\hat{H}$ is the Hamiltonian operator representing the total energy of the system. We shall initially consider the case where $\hat{H}$ is not itself explicitly time dependent and proceed to the more general case later, when we shall consider examples where $\hat{H}$ changes suddenly from one time-independent form to another and also where the time dependence is confined to a small part of the Hamiltonian when perturbation theory can be used. This will enable us to solve the problem of an atom subject to a time-varying field and will lead to an understanding of the occurrence of transitions between different energy states and of why some of these are more probable, and therefore associated with more intense spectral lines, than others. We shall then discuss the Ehrenfest theorem which is used to clarify the connection between quantum and classical mechanics. We close the chapter with a discussion of the application of the general results to the particular example of the ammonia maser.

8.1 TIME-INDEPENDENT HAMILTONIANS

This case was considered briefly in the discussion following Postulate 4.5 where it was shown that, if the wave function at some initial time $t = 0$ is given by $\Psi(\mathbf{r}, 0)$ where

$$\Psi(\mathbf{r}, 0) = \sum_n a_n(0)u_n(\mathbf{r}) \tag{8.2}$$

and the $u_n(\mathbf{r})$ are the energy eigenfunctions, that is,

$$\hat{H}u_n = E_n u_n \tag{8.3}$$

then the wave function $\Psi(\mathbf{r}, t)$ at time t is given by

$$\Psi(\mathbf{r}, t) = \sum_n a_n(0)u_n(\mathbf{r}) \exp(-iE_n t/\hbar) \tag{8.4}$$

We pointed out that (8.4) contains the quantum-mechanical equivalent of the conservation of energy: if the energy has been once measured so that the wave function is then an energy eigenfunction, it retains this form indefinitely (apart from a time-dependent phase factor) and any subsequent energy measurement is then certain to produce an identical result.

Equation (8.4) can also be used to study the behaviour of systems that are not initially in energy eigenstates, and we shall now consider two particular examples which illustrate this point.

The Harmonic Oscillator

The energy levels of a one-dimensional harmonic oscillator were shown in Chapter 2 to be $E_n = (n + \frac{1}{2})\hbar\omega$ where ω is the classical angular frequency. We can use this expression and the general theory described above to study the behaviour of such an oscillator when it is not initially in an energy eigenstate. Equation (8.4) becomes, in this case,

$$\Psi(x, t) = \sum_n a_n(0)u_n(x) e^{-i(n + 1/2)\omega t} \tag{8.5}$$

If we consider the particular time T corresponding to the classical period of the oscillator (that is, $T = 2\pi/\omega$) we get

$$\Psi(x, T) = \sum_n a_n(0)u_n(x) e^{-i(n + 1/2)2\pi}$$

$$= -\sum_n a_n(0)u_n(x)$$

$$= -\Psi(x, 0) \tag{8.6}$$

Thus, whatever the initial conditions, the wave function at time T is equal to minus that at time zero. It clearly follows that at time $2T$ the wave function will be identical to that at time zero, so the wave function varies periodically with a frequency half that of the classical oscillator. However, it should be noted that the sign of the wave function has no physical significance, and we can therefore conclude that all the physical properties of the harmonic oscillator will vary periodically in time with a frequency identical to the classical frequency. In particular, if we were to observe the position of the particle at time $t = 0$ so that its wave function was then strongly localized near a particular place, we should expect to find this form reproduced and the particle therefore in the same region, at times T, $2T$, $3T$, etc. Figure 8.1 illustrates this by showing the time evolution of the position probability density associated with a wave function whose form at $t = 0$ is the normalized sum of the two lowest energy eigenfunctions.

We can use this example as a further illustration of the correspondence principle, previously discussed in connection with the harmonic oscillator in Chapter 2. This states that the results of quantum mechanics go over to those of classical mechanics in the classical limit where the total energy is large compared with the separation of the states. A classical oscillator, such as a clock pendulum, appears always to have a well-defined position which

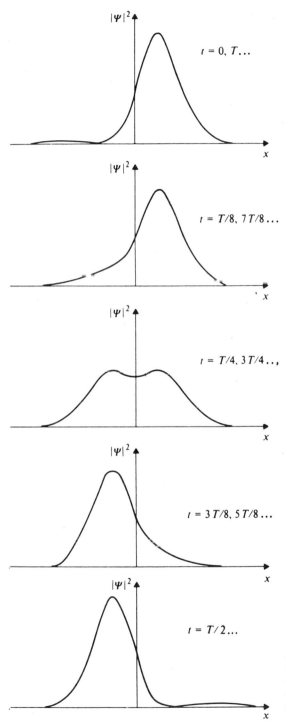

$t = 0, T \ldots$

$t = T/8, 7T/8 \ldots$

$t = T/4, 3T/4 \ldots,$

$t = 3T/8, 5T/8 \ldots$

$t = T/2 \ldots$

Figure 8.1 The time evolution of the position probability density corresponding to a particle in a harmonic oscillator potential whose wave function at $t = 0$ is the normalized sum of the two lowest energy eigenfunctions.

varies sinusoidally with the classical frequency; quantum mechanically this means that the initial wave function must have the form of a narrow localized pulse, which moves from side to side with the classical frequency and amplitude. This implies that a large number of energy eigenfunctions must contribute to the expansion (8.5). There must therefore be some uncertainty in the value of the energy, and the classical limit is reached when the average energy is so large that this uncertainty is an undetectably small fraction of the total.

Spin-Half Particle in a Magnetic Field

Although the time-dependent Schrödinger equation (8.1) has been written in terms of the wave representation, it is equally valid if Ψ is replaced by a column vector whose elements are in general time dependent, and the Hamiltonian is expressed as a matrix (cf. Chapter 6). If we consider the particular case of a spin-half particle with no orbital angular momentum in a magnetic field B which is in the z direction, (8.1) becomes

$$i\hbar \frac{\partial}{\partial t} \begin{bmatrix} a_1(t) \\ a_2(t) \end{bmatrix} = \hat{H} \begin{bmatrix} a_1(t) \\ a_2(t) \end{bmatrix} \tag{8.7}$$

and the Hamiltonian $\hat{H}$ is given by (6.47) as

$$\hat{H} = \frac{e}{m_e} B \hat{S}_z$$

$$= \frac{e\hbar}{2m_e} B \begin{bmatrix} 1 & 0 \\ 0 & -1 \end{bmatrix} \tag{8.8}$$

where $\hat{S}_z$ has been expressed in terms of a spin-matrix (cf. (6.15) and (6.16)). Substituting from (8.8) into (8.7) and expanding we get

$$\left. \begin{aligned} i\hbar \frac{\partial a_1}{\partial t} &= (e\hbar B/2m_e)a_1 \\ \\ i\hbar \frac{\partial a_2}{\partial t} &= -(e\hbar B/2m_e)a_2 \end{aligned} \right\} \tag{8.9}$$

and therefore

$$\left. \begin{aligned} a_1(t) &= a_1(0)\, e^{-(1/2)i\omega_p t} \\ a_2(t) &= a_2(0)\, e^{(1/2)i\omega_p t} \end{aligned} \right\} \tag{8.10}$$

where $\omega_p = eB/m_e$. If the initial state is an eigenstate of the energy (and hence of $\hat{S}_z$) at $t = 0$, then $a_1(0) = 1$ and $a_2(0) = 0$ (or vice versa depending on whether the eigenvalue is plus or minus $\frac{1}{2}\hbar$) and the only change in time is that contained in a phase factor multiplying the whole state vector which has no physical consequences. However, a more interesting case is where the initial state is an eigenstate of $\hat{S}_x$: for example, that where $a_1(0) = a_2(0) = 2^{-1/2}$, corresponding to an eigenvalue of $\frac{1}{2}\hbar$. It follows directly from (8.10) that the wave vector at time t now has the form

$$\frac{1}{\sqrt{2}}\begin{bmatrix} e^{-(1/2)i\omega_p t} \\ e^{(1/2)i\omega_p t} \end{bmatrix} \tag{8.11}$$

This expression is again an eigenvector of $\hat{S}_x$ with the same eigenvalue when $t = 2\pi/\omega_p$, $4\pi/\omega_p$, $6\pi/\omega_p$, etc. Moreover, we can easily show that at other times (8.11) is an eigenvector of the operator $\hat{S}_\phi$ representing a measurement of the component of spin in a direction in the xy plane at an angle ϕ to the x axis where $\phi = \omega_p t$. This is proved by considering the matrix representing the operator $\hat{S}_\phi$:

$$\hat{S}_\phi = \hat{S}_x \cos\phi + \hat{S}_y \sin\phi$$

$$= \tfrac{1}{2}\hbar \begin{bmatrix} 0 & e^{-i\phi} \\ e^{i\phi} & 0 \end{bmatrix} \tag{8.12}$$

The eigenvector of this matrix with eigenvalue $\frac{1}{2}\hbar$ is easily seen to be

$$\frac{1}{\sqrt{2}}\begin{bmatrix} e^{-i\phi/2} \\ e^{i\phi/2} \end{bmatrix} \tag{8.13}$$

and the required result follows directly from a comparison of (8.11) and (8.13). We conclude, therefore, that if the system is initially in an eigenstate of $\hat{S}_x$ with eigenvalue $\frac{1}{2}\hbar$, it will always be in a similar eigenstate of the operator $\hat{S}_\phi$ whose direction rotates in the xy plane with an angular velocity ω_p. It is tempting to conclude from this that the angular-momentum vector of the particle precesses about the field direction with this angular velocity, which is what would happen in the similar classical situation. But it is important not to pursue this analogy too far. In classical precession, the direction of the angular-momentum vector, and hence the magnitudes of all three of its components, always have known values; but in quantum mechanics only one angular-momentum component can be measured at any given time. For example, the precession model would imply that the y component of the angular momentum would be zero at the times when ϕ is zero, but we know from quantum mechanics and experiment that a measurement of this quantity always yields a result equal to either plus or minus $\frac{1}{2}\hbar$ and never zero. The application of this precession model (sometimes known as the 'vector model') to a quantum-mechanical system therefore incorrectly attributes properties to the system that are additional to those predicted by quantum mechanics; theories of this types are known as 'hidden-variable' theories and will be discussed in more detail in Chapter 11.

Finally we note a further interesting consequence of Eq. (8.13): if we add 2π to the angle ϕ, that is if we rotate the system through 360°, the sign of the eigenfunction is reversed, and we see from (8.11) that this sign change also occurs under the influence of a magnetic field after a time $2\pi/\omega_p$. Usually this has no effect on the physical properties of the system, for the same reasons as were discussed above in connection with the sign of the wave function of the harmonic oscillator. However, experiments can be conducted in which the occurrence of the sign change is confirmed: a beam of spin-half particles is directed into an apparatus through which there are two possible paths (see Fig. 8.2); each path contains a region in which there is a magnetic field and the two magnetic fields have the same magnitude, but are in opposite

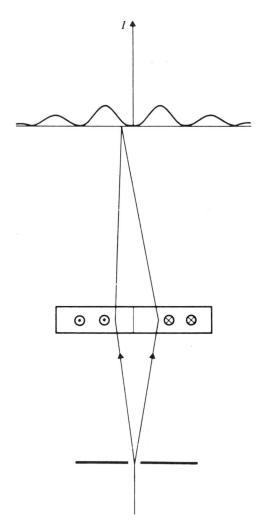

Figure 8.2 Neutrons pass through the lower slit and then enter a region where there is a magnetic field directed either vertically upwards (on left) or downwards (on right). The resulting interference pattern has a minimum in the centre if the relative rotation of the spins passing along the two paths is an odd number times 2π.

directions; the speed of the particles is adjusted so that the magnetic fields would rotate the spin by 180° in opposite directions depending on the path followed, and the wave functions associated with each path would therefore have opposite sign. An interference experiment is performed in which we do not know which route the particles have followed and the wave function on the far side of the apparatus is therefore the sum of those associated with the two paths; the resulting diffraction pattern therefore has zero intensity at its centre due to the destructive interference between the oppositely rotated components. Such experiments were carried out by a number of workers in the mid 1970s using neutrons, and the results were in complete agreement with theoretical predictions.

8.2 THE SUDDEN APPROXIMATION

So far we have restricted our discussion to systems whose Hamiltonians have no explicit time dependence, but we shall now extend our treatment to include cases where time-varying forces are acting. In general these problems are very difficult to solve and we shall restrict our consideration to those where particular simplifying assumptions can be applied. One such case is the *sudden approximation* which can be used when the Hamiltonian changes instantaneously from one time-independent form—say $\hat{H}_1$—to another—say $\hat{H}_2$—at a time which we take to be $t = 0$. Thus

$$\left.\begin{aligned} \hat{H} &= \hat{H}_1 \qquad t \leqslant 0 \\ \hat{H} &= \hat{H}_2 \qquad t > 0 \end{aligned}\right\} \tag{8.14}$$

We assume that the eigenfunctions of $\hat{H}_1$ and $\hat{H}_2$ are u_n and v_n respectively and that the system is known to be in one of the eigenstates of $\hat{H}_1$—say that represented by u_k before the change. We shall obtain the form of the wave function at times $t > 0$ and hence the probabilities that a subsequent energy measurement will yield a particular eigenvalue of $\hat{H}_2$.

The solution to this problem is obtained quite straightforwardly using the fact that ψ must be continuous in time, even when $\hat{H}$ changes discontinuously as at $t = 0$ (cf. Postulate 4.1). Thus immediately before and after the change we must have

$$\Psi(\mathbf{r}, 0) = u_k(\mathbf{r})$$
$$= \sum_n a_n(0)v_n(\mathbf{r}) \tag{8.15}$$

where we have used completeness to expand u_k in terms of the set of eigenfunctions v_n. As $\hat{H}_2$ is time independent we can use (8.4) to obtain an expression for Ψ at all times greater than zero:

$$\Psi(\mathbf{r}, t) = \sum_n a_n(0)v_n(\mathbf{r}) \exp(-iE_n t/\hbar) \tag{8.16}$$

where the energy levels E_n are the eigenvalues of $\hat{H}_2$. Expressions for the constants $a_n(0)$ can be obtained by multiplying both sides of (8.15) by v_n^* and integrating over all space. Thus

$$a_n(0) = \int v_n^* u_k \, d\tau \tag{8.17}$$

According to the quantum theory of measurement, the probability of obtaining any particular value E_n as a result of a measurement of the energy at any time after the change is equal to $|a_n|^2$. Following such a measurement, of course, the wave function would be changed to equal the corresponding eigenfunction v_n.

An example of the practical application of the sudden approximation is the change in the wave function of an atom following a radioactive decay of its nucleus. Tritium (^{3}H) can decay by the emission of a β particle and a neutrino to become a positively charged, one-electron ion whose nucleus is ^{3}He. As far as the atomic electron is concerned, therefore, its Hamiltonian has changed suddenly from that corresponding to a hydrogen atom with nuclear charge $Z = 1$ to that of a He$^+$ ion with $Z = 2$. The energy eigenfunctions of both these systems can be obtained from the expressions given in Chapter 3 and the probabilities of subsequent measurements yielding particular eigenvalues of the He$^+$ ion can then be readily calculated following the procedure described above.

A particularly interesting feature of the above example follows from the fact that a value of the energy of the He$^+$ ion can in principle be obtained from a knowledge of the energy associated with the nuclear decay, combined with those of the emitted β particle and the neutrino (although in practice the energy of the latter would be very difficult to measure). But the β particle and neutrino could well be a large distance from the atom when these measurements are made so that the energy of the ion would have been measured without apparently interfering with it. Nevertheless, quantum mechanics states that this measurement will cause the wave function of the atom to change from a form similar to (8.16) to the appropriate energy eigenfunction. This apparent contradiction is an example of the Einstein–Podolski–Rosen paradox which will be discussed more fully in Chapter 11.

8.3 TIME-DEPENDENT PERTURBATION THEORY

A very important type of time-dependent problem is one where the Hamiltonian $\hat{H}$ can be written as the sum of a time-independent part $\hat{H}_0$ and a small time-dependent perturbation $\hat{H}'$. An example of this, to which we shall return later, is the case of an atom subject to an oscillating electric field, such as that associated with an electromagnetic wave which can cause transitions to occur from one energy state to another. We shall now describe a method for obtaining approximate solutions to such problems which is known as time-dependent perturbation theory. It is similar, but of course not identical, to time-independent perturbation theory which was discussed in Chapter 7.

We wish to solve the time-dependent Schrödinger equation (8.1) for the case where

$$\hat{H}(\mathbf{r}, t) = \hat{H}_0(\mathbf{r}) + \hat{H}'(\mathbf{r}, t) \tag{8.18}$$

We assume that the eigenfunctions $u_k(\mathbf{r})$ of $\hat{H}_0$ are known and expand the wave function $\Psi(\mathbf{r}, t)$ as a linear combination of these

$$\Psi(\mathbf{r}, t) = \sum_k c_k(t) u_k(\mathbf{r}) e^{-iE_k t/\hbar} \tag{8.19}$$

where the expansion coefficients c_k have been defined so as to exclude the factors $\exp(-iE_k t/\hbar)$, as this simplifies the ensuing argument. Substituting from (8.18) and (8.19) into (8.1) we get

$$i\hbar \sum_k (\dot{c}_k - i\omega_k c_k) u_k \, e^{-i\omega_k t} = \sum_k (c_k \hbar \omega_k u_k \, e^{-i\omega_k t} + c_k \hat{H}' u_k \, e^{-i\omega_k t})$$

where

$$\dot{c}_k = \frac{\partial c_k}{\partial t} \quad \text{and} \quad \omega_k = E_k/\hbar$$

Thus

$$\sum_k (i\hbar \dot{c}_k - c_k \hat{H}') u_k \, e^{-i\omega_k t} = 0 \tag{8.20}$$

We now multiply (8.20) by u_m^* and integrate over all space to get

$$i\hbar \dot{c}_m \, e^{-i\omega_m t} - \sum_k c_k \hat{H}'_{mk} \, e^{-i\omega_k t} = 0$$

That is

$$\dot{c}_m = \frac{1}{i\hbar} \sum_k c_k \hat{H}'_{mk} \, e^{i\omega_{mk} t} \tag{8.21}$$

where

$$\hat{H}'_{mk} = \int u_m^* \hat{H}' u_k \, d\tau \quad \text{and} \quad \omega_{mk} = \omega_m - \omega_k$$

Everything done so far is exact, but we now apply perturbation techniques in a similar manner to that described for the time-independent case in Chapter 7. We introduce a constant β, replacing $\hat{H}'$ by $\beta \hat{H}'$ and expanding the constants c_k in a perturbation series

$$c_k = c_{k0} + \beta c_{k1} \ldots \tag{8.22}$$

Substituting from (8.22) into (8.21), remembering to replace H'_{mk} by $\beta H'_{mk}$ we get

$$\dot{c}_{m0} + \beta \dot{c}_{m1} + \cdots = \frac{1}{i\hbar} \beta \sum_k c_{k0} H'_{mk} \, e^{i\omega_{mk} t} + \cdots \tag{8.23}$$

where the omitted terms all contain higher-order powers of β. Equating the coefficients of the zeroth and first powers of β we have

$$\dot{c}_{m0} = 0 \qquad (8.24)$$

$$\dot{c}_{m1} = \frac{1}{i\hbar} \sum_k c_{k0} H'_{mk} e^{i\omega_{mk}t} \qquad (8.25)$$

Equation (8.24) implies that the coefficients c_{m0} are constant in time, which is to be expected as the zero-order Hamiltonian is time independent. The first-order contributions to c_m are obtained from (8.25) as

$$c_{m1} = \frac{1}{i\hbar} \sum_k c_{k0} \int_0^t H'_{mk} e^{i\omega_{mk}t} \, dt \qquad (8.26)$$

We are particularly interested in the case where the system is known to be in a particular eigenstate—say that represented by u_n—at the time $t = 0$ so that $c_{n0} = 1$, and $c_{k0} = 0$, $k \neq n$. Equation (8.26) then becomes

$$c_{m1} = \frac{1}{i\hbar} \int_0^t H'_{mn} e^{i\omega_{mn}t} \, dt \qquad (8.27)$$

Remembering that $c_{m0} = 0$ $(m \neq n)$ we see that the probability of finding the system in a state represented by u_m where $m \neq n$ is given by $|c_{m1}|^2$.

Periodic Perturbations

We shall now apply these general results to systems where the perturbation varies sinusoidally in time, that is, where

$$\hat{H}'(\mathbf{r}, t) = \hat{H}''(\mathbf{r}) \cos \omega t \qquad (8.28)$$

and ω is the angular frequency of the perturbation. Writing the cosine as a sum of two exponentials and substituting into (8.27), we get

$$
\begin{aligned}
c_{m1} &= \frac{H''_{mn}}{2i\hbar} \int_0^t \left[e^{i(-\omega + \omega_{mn})t} + e^{i(\omega + \omega_{mn})t} \right] dt \\
&= -\frac{H''_{mn}}{2\hbar} \left[\frac{e^{i(-\omega + \omega_{mn})t} - 1}{-\omega + \omega_{mn}} + \frac{e^{i(\omega + \omega_{mn})t} - 1}{\omega + \omega_{mn}} \right]
\end{aligned} \qquad (8.29)
$$

If we now consider the first term in square brackets on the right-hand side of (8.29) we see that this has a maximum value if $\omega = \omega_{mn}$ and is comparatively small for values of ω appreciably different from this, while the second term has a similar maximum at $\omega = -\omega_{mn}$. It is therefore a reasonable approximation to assume that c_{m1}, and therefore the transition probability, is negligibly small unless one of these conditions is, at least approximately, fulfilled; moreover (except for the special case where $\omega_{mn} = 0$, to which we shall return in the next chapter) both conditions cannot be simultaneously satisfied, so we can assume that only one of the two terms is non-negligible for a particular value of ω. We shall consider the case where $\omega \simeq \omega_{mn}$ and neglect the second term in (8.29) for the moment.

The probability of finding the system in the state u_m at time t is therefore $|c_{m1}|^2$ where

$$|c_{m1}|^2 = \frac{|H''_{mn}|^2}{4\hbar^2} \frac{2[1 - \cos(\omega_{mn} - \omega)t]}{(\omega_{mn} - \omega)^2}$$

$$= \frac{|H''_{mn}|^2}{4\hbar^2} \frac{\sin^2[(\omega_{mn} - \omega)t/2]}{[(\omega_{mn} - \omega)/2]^2} \qquad (8.30)$$

This expression is plotted as a function of $(\omega_{mn} - \omega)$ in Fig. 8.3. We note that the height of the central peak is proportional to t^2 while its width is proportional to t^{-1}. Thus, after a time that is long compared with the period of the perturbation, the transition probability will be negligibly small unless the condition $\omega = \omega_{mn}$ is fulfilled. That is, unless

$$\hbar\omega = E_m - E_n \qquad (8.31)$$

In the case where the perturbation results from an electromagnetic wave, ω is the angular frequency of the radiation so we see that (8.31) is just the

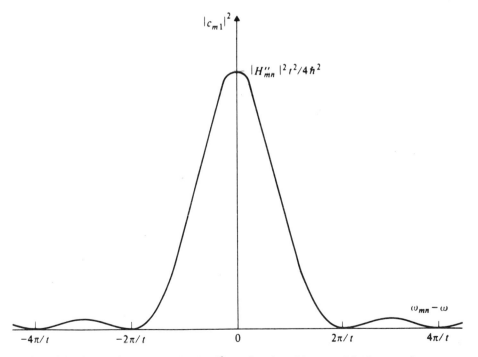

Figure 8.3 The transition probability $|c_{m1}|^2$ as a function of $(\omega_{mn} - \omega)$ in the case of a system which has been subject to a periodic perturbation of angular frequency ω for a time t.

basic equation relating this quantity to the difference between the energies of the states which we discussed in Chapter 1. It follows that, if we were to consider the perturbing wave to be composed of photons, (8.30) would represent the probability that a photon of energy $\hbar\omega_{mn}$ had interacted with the atom within the time t; we should note, however, that the quantization of the field in terms of photons is not necessary to explain the phenomena described so far in this chapter.

We can draw another conclusion from (8.31) by noting that, since ω is always positive, E_m must be larger than E_n and the transition discussed corresponds to an excitation of the system, with, presumably, a corresponding absorption of energy from the perturbing field. If, however, we had used the second instead of the first term on the right-hand side of (8.29), we should have obtained the condition $\omega = -\omega_{mn}$ along with expressions similar to (8.30) and (8.31), but with appropriate changes of sign. This would then correspond to the case where the perturbation causes a transition from an initial excited state to another of lower energy; energy is then emitted from the system and such a process is known as 'stimulated emission'.

Before we can use (8.30) to calculate more detailed properties of transitions in practical situations, such as the absorption or emission of electromagnetic radiation by atoms in gases, we must allow for the fact that (except in particular situations such as lasers which we refer to below) the excited states of quantum systems do not have perfectly defined energies, but are usually broadened into a band containing a large number of closely spaced levels. Furthermore, the perturbation is often not a pure oscillation at a single frequency ω, but is rather a mixture of frequencies in a band of greater or lesser width centred on ω. We first consider the case where the broadening of the energy levels is much greater than that of the perturbation. This broadening can arise from a number of causes: for example the atoms in a gas are in thermal motion so, relative to their own frame of reference, they 'see' the frequency of the perturbation shifted to a greater or lesser extent by the Doppler effect: in the laboratory frame of reference this is equivalent to a broadening of the atomic energy levels. Another cause of broadening, to be discussed in more detail shortly, arises because the excited states of atoms always have a 'natural' line width associated with the possibility of spontaneous emission. If one or both of the levels E_m and E_n are broadened for any reason, this results in a similar broadening of the frequency difference ω_{mn}. The latter can then be represented by a function $g(\omega_{mn})$, known as the 'density of states', which is defined so that the number of pairs of levels which have an energy difference between $\hbar\omega_{mn}$ and $\hbar(\omega_{mn} + d\omega_{mn})$ is $g(\omega_{mn})\,d\omega_{mn}$. Assuming that the matrix elements are identical for all these pairs of states, the total probability $P(t)$ for any such transition to take place is then

$$P(t) = \frac{|H''_{mn}|^2}{4\hbar^2} \int \frac{\sin^2[(\omega_{mn} - \omega)t/2]}{[(\omega_{mn} - \omega)/2]^2} g(\omega_{mn})\,d\omega_{mn} \qquad (8.32)$$

For large enough t, $g(\omega_{mn})$ is a much more slowly varying function of ω_{mn} than is the rest of the integrand, which is sharply peaked about the point $\omega_{mn} = \omega$ (cf. Fig. 8.3); we can therefore replace $g(\omega_{mn})$ by $g(\omega)$ and take this quantity outside the integral, which can then be evaluated by standard techniques, leading to

$$P(t) = \frac{\pi |H''_{mn}|^2}{2\hbar^2} g(\omega) t \qquad (8.33)$$

The transition rate, W, defined as the transition probability per unit time is then given by

$$W = \frac{dP}{dt} = \frac{\pi |H''_{mn}|^2}{2\hbar^2} g(\omega) \qquad (8.34)$$

In the case where the perturbation has a density of states $g'(\omega)$ which is much broader than the width of the levels, a similar argument leads to the same expression, but with $g(\omega)$ replaced by $g'(\omega_{mn})$. In the general case where both widths have to be taken into account, the relevant quantity is $\int g(\omega) g'(\omega) \, d\omega$.

Equation (8.34) is known as *Fermi's golden rule*. It has a wide range of application, both in the field of atomic transitions and, as we shall see in the next chapter, in scattering theory. The important point to remember is that the transition rate per unit time is proportional to the square of the matrix element and to the density of states.

Although energy levels are normally broadened sufficiently for Fermi's golden rule to be applicable, there are some situations in which this is not the case. In a typical laser, for example, the atoms in the lasing medium do not operate independently, but have to be treated as a coherent whole whose transition frequencies are determined by the size and shape of the laser cavity, and the resulting densities of states can be very narrow indeed. As a result, the transition probability shown in Fig. 8.2 does not in this case become narrower than $g(\omega_{mn})$ until t is so large that the value of $|c_{m1}|^2$ at $\omega_{mn} = \omega$ is greater than one. This unreasonable result arises because first-order perturbation theory does not allow for the fact that the probability of occupation of the original state u_n decreases as that relating to u_m rises. A more rigorous analysis, which we shall not describe here, shows that such a system oscillates between the two states at an angular frequency equal to $|H''_{mn}|/2\hbar$. Such 'quantum oscillations' play an important role in the behaviour of lasers and masers, as we shall see in the last section of this chapter where we describe the ammonia maser.

8.4 SELECTION RULES

The results of the previous section will now be used to improve our understanding of the absorption and emission of radiation by atoms undergoing transitions between energy levels. The transition rate clearly determines the intensity of the corresponding spectral line, and both are proportional to the square of the matrix element H''_{mn}. We note in particular that if this is zero the transition will not take place. Such non-occurring transitions are described as 'forbidden' and the rules that determine which transitions in a given system are forbidden and which are 'allowed' are known as *selection rules*. These are very important in the study of the physics of atoms, molecules, nuclei, and solids, and are extensively discussed in textbooks specializing in these areas. We shall confine our discussion to a brief introduction to the selection rules applying in the case of a one-electron atom.

We consider such an atom subject to a plane electromagnetic wave of angular frequency ω whose electric vector is in the z direction and has magnitude F where

$$F = F_0 \cos(\mathbf{k} \cdot \mathbf{r} - \omega t) \tag{8.35}$$

We take the origin of coordinates at the atomic nucleus and assume that the wavelength of the radiation is much greater than the radius of the atom. The term $\mathbf{k} \cdot \mathbf{r}$ is therefore very small so that the electric field is effectively uniform over the atom. The operator representing the energy of interaction between the atom and the field is therefore given by

$$\hat{H}' = eFz \simeq eF_0 z \cos \omega t \tag{8.36}$$

and therefore

$$\hat{H}'' = eF_0 z$$

We now consider the matrix element of this perturbation which connects two states whose quantum numbers are (n_1, l_1, m_1) and (n_2, l_2, m_2) respectively and which we write as H''_{12} for convenience. Using expressions for the one-electron wave functions from Chapter 3 (3.70), we have

$$H''_{12} = eF_0 \int u^*_{n_1 l_1 m_1} z u_{n_2 l_2 m_2} \, d\tau$$

$$\propto eF_0 \int_0^{2\pi} \int_0^{\pi} \int_0^{\infty} R^*_{n_1 l_1}(r) P_{l_1}^{|m_1|}(\cos\theta) e^{-im_1\phi} r \cos\theta$$

$$\times R_{n_2 l_2}(r) P_{l_2}^{|m_2|}(\cos\theta) e^{im_2\phi} r^2 \, dr \sin\theta \, d\theta \, d\phi$$

$$= eF_0 \left[\int_0^{\infty} R^*_{n_1 l_1}(r) R_{n_2 l_2}(r) r^3 \, dr \right]$$

$$\times \left[\int_0^{\pi} P_{l_1}^{|m_1|}(\cos\theta) P_{l_2}^{|m_2|}(\cos\theta) \cos\theta \sin\theta \, d\theta \right]$$

$$\times \left[\int_0^{2\pi} e^{i(m_2 - m_1)\phi} \, d\phi \right] \tag{8.37}$$

It follows that such a transition can take place only if all three integrals in (8.37) are non-zero. If we first consider the integral over ϕ, we see that this vanishes unless $m_1 = m_2$. The integral with respect to θ can be evaluated, given the following property of the associated Legendre functions, $P_l^{|m|}$:

$$(2l + 1) \cos \theta P_l^{|m|} = (l - |m| + 1)P_{l+1}^{|m|} + (l + |m|)P_{l-1}^{|m|} \qquad (8.38)$$

Using (8.38) and considering the case where $m_1 = m_2$ the integral over θ in (8.37) becomes

$$\frac{l_2 - |m_1| + 1}{2l_2 + 1} \int_0^\pi P_{l_1}^{|m_1|} P_{l_2+1}^{|m_1|} \sin \theta \, d\theta + \frac{l_2 + |m_1|}{2l_2 + 1} \int_0^\pi P_{l_1}^{|m_1|} P_{l_2-1}^{|m_1|} \sin \theta \, d\theta \qquad (8.39)$$

However, associated Legendre functions that have the same value of m but different l are orthogonal, so the integrals in (8.39) vanish unless $l_1 = l_2 + 1$ or $l_1 = l_2 - 1$. So far we have discussed a field polarized in the z direction; if we now consider it to be polarized in the x direction we will have

$$\hat{H}'' = eF_0 x$$

$$= eF_0 r \sin \theta \cos \phi \qquad (8.40)$$

in spherical polar coordinates. A similar argument to the above shows that in this case the same relation holds between l_1 and l_2, but that the condition on m_1 and m_2 is now

$$m_1 = m_2 \pm 1 \qquad (8.41)$$

The selection rules governing the allowed transitions in the presence of an electromagnetic wave of arbitrary polarization are therefore

$$\left. \begin{array}{l} \Delta l = \pm 1 \\ \Delta m = \pm 1 \text{ or } 0 \end{array} \right\} \qquad (8.42)$$

where Δl and Δm represent the differences between the values of l and m associated with the two states. We note that there are no such rules governing the change in the principal quantum number n, which results from the fact that the first integral in (8.37) is always finite; its value can, however, vary considerably from one transition to another resulting in a corresponding variation in the intensities of the spectral lines.

The above selection rules refer to *electric dipole transitions* where the perturbing Hamiltonian has the form (8.36) or (8.40). However, other types of transition can be induced by electromagnetic radiation. Firstly, $(\mathbf{k} \cdot \mathbf{r})$ can be assumed to be small rather than zero, and the right-hand side of (8.35) can be expanded as a power series leading to

$$F = F_0[\cos \omega t + (\mathbf{k} \cdot \mathbf{r}) \sin \omega t - \tfrac{1}{2}(\mathbf{k} \cdot \mathbf{r})^2 \cos \omega t + \cdots] \qquad (8.43)$$

Only the first term of (8.43) has been considered so far and the others can lead to transitions known as *electric quadrupole, electric octopole*, etc.

Secondly, the atomic electron interacts with the magnetic field associated with the wave as well as the electric field. This leads to the possibility of transitions known as *magnetic dipole, magnetic quadrupole,* etc. Detailed calculations of the transition probabilities associated with both these classes of transition are most conveniently performed if the electromagnetic wave is expressed in vector-potential form. Details of this and the resulting selection rules can be found in textbooks on atomic physics: in general the spectral lines associated with allowed electric dipole transitions are much more intense than any others.

Spontaneous Emission

So far we have discussed only the case where transitions between energy levels are caused by the perturbation associated with an applied field. However, it is well known that an atom in an excited state will decay to its ground state, emitting radiation, even if it is completely isolated. This appears to contradict our previous statement that a system in an energy eigenstate (such as, presumably, an atom in an excited state) should remain in this state indefinitely. The resolution of this apparent paradox lies in the fact that we have ignored the quantization of the electromagnetic field: the Hamiltonian (8.28) is evaluated assuming the classical form of the electromagnetic wave, neglecting the fact that it is really quantized and consists of photons, and indeed a similar approximation was made in deriving the energy eigenvalue in the first place. A proper discussion of field quantization is well outside the scope of this book, but the following simplified argument explains its application to the present problem.

Classically, an electromagnetic wave consists of perpendicular electric and magnetic fields that oscillate at some angular frequency ω. The energy of the wave is divided equally between the electric and magnetic fields and equals $\varepsilon_0 \langle F^2 \rangle V$ where $\langle F^2 \rangle$ is the mean square amplitude of the electric field and V is the volume occupied by the wave. If we now assume that the field can be quantized in the same way as a mechanical oscillator (cf. Sec. 2.6), then the lowest energy it can have is $\frac{1}{2}\hbar\omega$. This means that, in the vacuum, where classically we would not expect any fields to exist, each possible mode of oscillation has this energy. The vacuum must therefore contain a fluctuating electric field (analogous to the zero-point motion of the oscillator) whose mean square amplitude is given by

$$\langle F^2 \rangle = \hbar\omega / 2\varepsilon_0 V \tag{8.44}$$

This field can cause an excited atom to 'spontaneously' decay with the emission of a photon. The reverse process is clearly impossible as energy must be conserved and the field energy is already at its minimum value. We can use the above along with our earlier results to calculate the expected rate of spontaneous emission from an excited state. To do this we need to know the density of states associated with the zero-point field. This is directly proportional to the number of modes of vibration in a volume V, which can be calculated by assuming the volume to be a square box of side L on the boundaries of which the electric field must be zero. It follows that the **k** vectors of the waves must satisfy the condition

$$\mathbf{k} = (k_x, k_y, k_z) = (n_1\pi/L, n_2\pi/L, n_3\pi/L) \tag{8.45}$$

where n_1, n_2 and n_3 are integers. There is therefore one mode in the volume π^3/V in 'k-space'. Waves whose wave vectors have magnitudes in the range k to $k + dk$ have angular frequencies between ck and $c(k + dk)$. These can point in any direction in the positive octant of k-space so the total number is

$$(\tfrac{1}{2}\pi\omega^2 \, d\omega/c^3)(V/\pi^3) = (V\omega^2/2\pi^2c^3) \, d\omega \tag{8.46}$$

This must equal $g'(\omega) \, d\omega$, where g' is the density of states, so we can combine this with (8.34) and (8.35) to calculate the transition rate as

$$W_{sp} = (e^2\omega^3/8\pi\varepsilon_0\hbar c^3)|z_{12}|^2 \tag{8.47}$$

We note that this rate is proportional to $|z_{12}|^2$ so spontaneous emission is subject to the same selection rules as stimulated emission and absorption discussed above.

8.5 THE EHRENFEST THEOREM

The time-dependent Schrödinger equation can be used to calculate the rate of change of the expectation value of a physical quantity

$$\frac{\partial\langle\hat{Q}\rangle}{\partial t} = \frac{\partial}{\partial t}\int \Psi^*\hat{Q}\Psi \, d\tau$$

$$= \int \left[\frac{\partial\Psi^*}{\partial t}\hat{Q}\Psi + \Psi^*\hat{Q}\frac{\partial\Psi}{\partial t} + \Psi^*\frac{\partial\hat{Q}}{\partial t}\Psi\right] d\tau$$

We can now substitute expressions for $\partial\Psi/\partial t$ and $\partial\Psi^*/\partial t$ obtained from (8.1) and its complex conjugate to get

$$\frac{\partial\langle\hat{Q}\rangle}{\partial t} = \int \frac{i}{\hbar}[(\hat{H}^*\Psi^*)\hat{Q}\Psi - \Psi^*\hat{Q}\hat{H}\Psi] \, d\tau + \left\langle\frac{\partial\hat{Q}}{\partial t}\right\rangle$$

$$= \frac{i}{\hbar}\int [\Psi^*(\hat{H}\hat{Q} - \hat{Q}\hat{H})\Psi] \, d\tau + \left\langle\frac{\partial\hat{Q}}{\partial t}\right\rangle$$

using the Hermitian property of $\hat{H}$. That is

$$\frac{\partial\langle\hat{Q}\rangle}{\partial t} = \frac{i}{\hbar}\langle[\hat{H}, \hat{Q}]\rangle + \left\langle\frac{\partial\hat{Q}}{\partial t}\right\rangle \tag{8.48}$$

where $[\hat{H}, \hat{Q}]$ is the commutator of $\hat{Q}$ and $\hat{H}$. Thus the rate of change of the expectation value of the physical quantity represented by $\hat{Q}$ is equal to a sum of two terms, the first of which is proportional to the expectation value of the commutator of $\hat{Q}$ with the Hamiltonian of the system, $\hat{H}$, while the second is equal to the expectation value of $\partial\hat{Q}/\partial t$. This result is known as the Ehrenfest theorem. We note that in the particular case where $\hat{Q}$ is time independent and commutes with $\hat{H}$, the expectation value of $\hat{Q}$ is constant, in agreement with the fact that the value of the corresponding physical quantity is conserved (cf. Sec. 4.6).

We can use the Ehrenfest theorem to investigate further the connection between quantum and classical mechanics. Assuming that in the classical limit the measured value of a physical quantity is indistinguishable from its quantum-mechanical expectation value, the laws of classical mechanics can be derived from (8.48). For example, the fact that energy is conserved in a closed system follows by putting $\hat{Q}$ equal to $\hat{H}$, remembering that in such a case the latter is time independent. Alternatively, if we take $\hat{Q}$ to be the x component of the momentum—that is, $\hat{Q} = \hat{P}_x = -i\hbar\partial/\partial x$—and express $\hat{H}$ as the sum of a kinetic energy term $(\hat{P}^2/2m)$ and a potential $V(r)$, (8.48) gives

$$\frac{\partial\langle\hat{P}_x\rangle}{\partial t} = \frac{1}{\hbar}\left\langle\left[\left(-\frac{\hbar^2}{2m}\nabla^2 + V\right), -i\hbar\frac{\partial}{\partial x}\right]\right\rangle$$

$$= -\left\langle\frac{\partial V}{\partial x}\right\rangle \tag{8.49}$$

because all other terms cancel when the commutator bracket is expanded. But, if $\langle P_x\rangle$ and $\langle\partial V/\partial x\rangle$ are equivalent to the classical limits of these quantities, (8.49) simply states that the rate of change in the x component of momentum is equal to the x component of the applied force, which is Newton's second law.

Thus the correspondence principle follows from the Ehrenfest theorem, provided we can identify the quantum-mechanical expectation value of a physical quantity with its value in classical mechanics. In most classical systems this condition is clearly satisfied: macroscopic particles are normally strongly localized and wave function spreads are therefore very narrow compared with the dimensions of the measuring apparatus.

However, there are some circumstances in which the results of quantum mechanics do not go over to those of classical mechanics in such a direct way. Consider, for example, the case of a particle passing through the two slits of a Young's interference experiment; its wave function will be finite over the area of each slit, but the expectation value of its position corresponds to a point midway between the slits where the wave function is zero and, however far apart the slits or however massive the particle, it is never observed at a position corresponding to this expectation value. What happens in this case is that the interference effects associated with the delocalization of the particle become harder and harder to observe in the classical limit and the system becomes indistinguishable from one in which the particle passes through either one of the two slits, but not both. Nevertheless, there is a theoretical possibility that a macroscopic system could be constructed which displayed quantum-mechanical delocalization and whose properties could not be accounted for classically. Some of the conceptual problems associated with such macroscopic quantum effects are discussed in Chapter 11.

8.6 THE AMMONIA MASER

A physical system that illustrates a number of the quantum-mechanical results discussed in this and previous chapters is the ammonia maser. A maser is a device that amplifies microwave radiation. It relies on the provision of a medium (in this case ammonia) which has more of its molecules in an excited state than there are in the ground state. Radiation of the correct frequency then induces transitions to the ground state and the consequent stimulated emission of more radiation. The way in which this process operates in the case of ammonia will be described in this section.

The ammonia molecule consists of a nitrogen atom bound chemically to three hydrogen atoms in such a way that the nitrogen atom lies a little above (or below) the centre of the equilateral triangle formed by the hydrogens (see Fig. 8.4). The exact wave function describing the four nuclei and ten electrons in the ammonia molecule is very complicated and would require considerable computational effort to evaluate—if indeed this were even possible. However, all we need to know for our present discussion is that there are two states which are equivalent apart from the position (up or down) of the nitrogen nucleus (see Fig. 8.4) and the fact that it is possible for the atom to pass from one position to the other by quantum-mechanical tunnelling.†

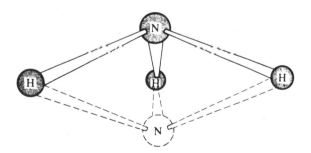

Figure 8.4 The geometrical configuration of the ammonia molecule in states where the nitrogen atom is above (continuous lines) and below (broken lines) the plane of the three hydrogens.

†It might be thought that the transition from the 'up state' to the 'down state' could be effected simply by rotating the molecule about an axis in the plane of the hydrogens. In fact in the relevant states the molecule possesses angular momentum about its symmetry axis which stabilizes its spatial orientation gyroscopically.

However, neither of the two states described so far is an energy eigenstate. This follows from the fact that the molecule is able to tunnel between the two configurations and that, as the potential is centrosymmetric, the energy eigenstates must have definite parity—that is, they must be either symmetric or antisymmetric with respect to reflection. If we call the wave functions corresponding to the nitrogen in the up and down positions ϕ_1 and ϕ_2 respectively, we can make linear combinations (u_1 and u_2) of these which do have definite parity and which are then reasonable approximations to the energy eigenfunctions

$$u_1 = \frac{1}{\sqrt{2}}(\phi_1 + \phi_2)$$

$$u_2 = \frac{1}{\sqrt{2}}(\phi_1 - \phi_2)$$

(8.50)

It follows that, if we know the position of the nitrogen atom relative to the hydrogens so fixing the wave function as ϕ_1 or ϕ_2, the molecule cannot be in a state of definite energy, whereas if we know the energy, the nitrogen position must be uncertain. We note that, although a nitrogen atom is not macroscopic, it is a lot heavier than an electron, and the fact that it can be delocalized is an interesting confirmation of the application of quantum mechanics to such an object.

The eigenfunction u_1 is symmetric and is taken to be the ground state of the system whose energy we represent by E_1, while u_2 is antisymmetric and is therefore an excited state (energy E_2). It turns out that the energy difference corresponds to a frequency in the microwave region and we should therefore be able to construct a maser provided that we can produce nitrogen molecules entirely, or at least predominantly, in the state u_2. Now, at normal temperatures, the average thermal energy of the molecules is much greater than the difference in energy between the two states, so if their energy were to be measured, an approximately equal number would be found in each state. In order to isolate a set of molecules all in the state u_2, we first create a molecular beam by allowing the gas to emerge from a container through a fine hole (see Fig. 8.5). We then measure the energy of each molecule, so placing it in one of the energy eigenstates. This is achieved by passing the beam through a region where there is a static non-uniform electric field, the effect of which we will now explain by considering the electrical polarizability of the ammonia molecule.

A general expression for the polarizability of a one-electron atom was obtained by perturbation theory in Chapter 7 (Eq. (7.28)). This can be applied to an ammonia molecule, provided the operator ez representing the instantaneous electric dipole moment of the atom is replaced by the corresponding molecular property. It is known that the electrons in the ammonia molecule are distributed in such a way that the nitrogen atom

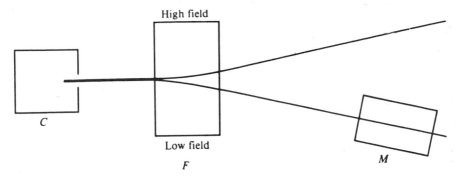

High field

Low field

C

F

M

Figure 8.5 The ammonia maser. Ammonia molecules leave the container *C* and pass through a region of non-uniform electric field *F*. Excited molecules are deviated towards the lower field region and directed into a microwave cavity *M*.

carries a net negative charge, which we denote by q, while a positive charge of equal magnitude is distributed between the three hydrogens. It follows that, if the nitrogen atom were at a distance x from the plane containing the hydrogens the molecule would have a dipole moment qx, and this quantity corresponds with ez in the atomic case. Moreover, the difference between the energies of the states u_1 and u_2 is much less than that between either of them and any other energy state of the system. These latter can therefore be ignored when considering the perturbation expression for the polarizabilities of the states which are therefore

$$\left.\begin{aligned}\alpha_1 &= \frac{2q^2|x_{12}|^2}{E_2 - E_1}\\\alpha_2 &= \frac{2q^2|x_{12}|^2}{E_1 - E_2}\end{aligned}\right\}\qquad(8.51)$$

where $x_{12} = \int u_1^* x u_2 \, d\tau$†. It follows that α_1 is positive while α_2 is negative, and the application of an electric field will induce a net dipole moment parallel to the field in the first case and anti-parallel in the second. The application of a non-uniform field therefore sorts the molecules into two groups: one, consisting of molecules in their ground state, moves to a region of high field intensity, while the other, containing only molecules in the excited state, moves to low-field regions. This process is closely analogous to the measurement of a spin component using a non-uniform magnetic field in the Stern–Gerlach experiment.

Having separated out a beam of excited molecules, the next step is to direct it into a microwave cavity whose dimensions are such that its natural frequency equals that associated with the transition. Microwave radiation

†Of course the field must be weak enough for the polarization energy to be much less than $E_2 - E_1$, otherwise the two energy eigenstates would be mixed as in the Stark effect.

of this frequency which enters the cavity will therefore stimulate the emission of more radiation resulting in the required amplification. In fact, the transition probability is typically so large and the line width in the cavity is so small, that quantum oscillations of the type described earlier in this chapter can occur: in one half-cycle of such an oscillation when the molecules are predominantly excited, amplification results, but in the other when the molecules are largely in the ground state, the incident radiation is absorbed. To avoid this, the speed of the molecular beam is chosen so that each molecule is in the cavity for a time approximately equal to half the period of the quantum oscillation; since on entry it is in its excited state, the absorption phase is thereby eliminated and continuous amplification results.

PROBLEMS

8.1 The wave function of a particle in a one-dimensional infinite-sided potential well of width $2a$ is $\psi = 2^{-1/2}(u_1 + u_2)$ at time $t = 0$ where u_1 and u_2 are the two lowest energy eigenfunctions. Find an expression for the position probability distribution as a function of time and show that it is periodic with angular frequency $\omega = 3\pi^2\hbar/8ma^2$. Sketch this probability distribution at times 0, $\pi/2\omega$, π/ω and $3\pi/2\omega$.

8.2 Show that the expectation value of the position of a particle in a harmonic oscillator potential oscillates sinusoidally with the classical frequency if the system is not in an energy eigenstate.

8.3 Show that the expectation value of the angular momentum of an electron in a magnetic field **B** precesses about the direction of **B** with an angular frequency eB/m_e, unless it is in an energy eigenstate.

8.4 What is the probability of finding the resulting $^3\mathrm{He}^+$ ion in (i) its $1s$, (ii) its $2s$, and (iii) one of its $2p$ states, following the β decay of a $^3\mathrm{H}$ atom initially in its ground state?

8.5 The spring constant of a harmonic oscillator in its ground state is suddenly doubled. Calculate the probability that a subsequent energy measurement will find the new oscillator in (i) its ground state, (ii) its first excited state, and (iii) its second excited state.

8.6 In the experiment on spin interference described in Sec. 8.1, the magnetic fields used had a magnitude of $0.5T$ and the path lengths were each 7×10^{-5} m. Show that a minimum in the centre of the diffraction pattern is to be expected for neutrons with a wavelength of 3.89×10^{-10} m, given that the neutron has a magnetic moment of magnitude 1.91 $(e\hbar/2m_n)$, m_n being the neutron mass.

8.7 A particle, initially in an energy eigenstate of an infinite-sided potential well, is subject to a perturbation of the form $V_0 x \cos \omega t$. Show that transitions are possible between the states u_n and u_m only if $n + m$ is odd.

8.8 The amplitude H'' associated with magnetic dipole transitions turns out to be proportional to the operator representing the angular momentum vector. Show that in a one-electron atom the selection rules for such transitions are $\Delta l = 0$, $\Delta m = \pm 1$ or 0. Which of these apply when the angular momentum vector is (i) parallel, and (ii) perpendicular to the z axis?

    Hint: Express L_x and L_y in terms of ladder operators.

8.9 Given that $yH_n = \frac{1}{2}H_{n+1} + nH_{n-1}$ where the H_n's are Hermite polynomials, show that the selection rule for electric dipole transitions in a one-dimensional harmonic oscillator is $\Delta n = \pm 1$.

8.10 A certain physical system has a Hamiltonian operator of the form $\hat{H}_0 + \hat{H}'' \cos \omega t$ where $\hat{H}_0$ and $\hat{H}''$ are time independent, but H'' need not be a small perturbation. The operator $\hat{H}_0$

has only two eigenstates whose eigenfunctions are u_1 and u_2 respectively. Show that the expression $au_1 \exp(iE_1t/\hbar) + bu_2 \exp(iE_2t/\hbar)$ is a solution to the time-dependent Schrödinger equation in this case if

$$\frac{da}{dt} = \frac{b}{i\hbar} H''_{21} \cos \omega t \, e^{i\omega_{21}t}$$

and

$$\frac{db}{dt} = \frac{a}{i\hbar} H''_{21} \cos \omega t \, e^{i\omega_{12}t}$$

in the usual notation, provided $H''_{11} = H''_{22} = 0$. Show that if $\omega = \omega_{12}$ and if high-frequency terms can be ignored, then

$$a = \cos(\Omega t - \phi) \qquad b = \sin(\Omega t - \phi)$$

are solutions to the above equations where $\Omega = |H''_{12}|/2\hbar$ and ϕ is a constant. Compare these results with the discussion of quantum oscillations in the text.

8.11 A particle moves in a one-dimensional potential well given by

$$V = V_0 \quad (-a \leqslant x \leqslant a) \qquad V = 0 \quad (a < |x| < b) \qquad V = \infty \quad (|x| > b)$$

Assuming that the magnitudes of a, b, and V_0 are such that the ground-state energy is less than V_0 and that there is a small, but finite, probability of tunnelling through the central barrier, draw sketch graphs of the two lowest energy eigenfunctions of this system. Discuss the evolution in time of such a system if the particle is known to be initially on one side of the barrier. Consider the response of this system to a perturbation whose frequency corresponds to the difference between the energies of the two lowest states and compare the properties of this system with those of the ammonia molecule.

NINE

SCATTERING

In this chapter we discuss the quantum mechanics of processes in which particles are scattered after colliding with a fixed object. Scattering is a very important feature of many physical processes and is nearly always a part of the process of performing and recording an experimental result. Thus when we perform an X-ray diffraction experiment we record the diffraction pattern created when X-rays are scattered by a crystal, while nearly all the information we possess about the energy levels and structure of nuclei and fundamental particles has been obtained from experiments in which beams of particles (e.g., protons or neutrons) are scattered from targets containing the nuclei or particles under investigation. Indeed any visual observation we make simply by looking at something involves the detection by our eyes of light scattered from the object we are looking at. It is clearly important therefore that we understand something of the quantum mechanics of the scattering processes, and this will be the aim of the present chapter.

We shall begin our study of the quantum mechanics of scattering by a consideration of a simple one-dimensional example which will illustrate a number of the points to be taken up in the subsequent discussion of three-dimensional systems. We shall introduce the latter by some general considerations; this will be followed by a description of the Born approximation in which the scattering process is treated as a weak perturbation, and we shall finally introduce the method known as partial wave analysis which is particularly useful in the description of the scattering of plane waves by spherical objects. Throughout this chapter we shall assume that the particles being scattered are of a different type from the object or particle doing the scattering, and we shall briefly discuss the particular features that arise in the case of collisions between identical particles towards the end of Chapter 10.

9.1 SCATTERING IN ONE DIMENSION

Scattering experiments are usually performed using beams of particles all initially moving with the same speed in the same direction, so we begin our study by considering the wave function of a particle moving with momentum p in the positive x direction. We know this must be an eigenfunction of the momentum operator $\hat{P}_x = -i\hbar\partial/\partial x$ so that

$$\psi_k(x) = A_k \exp(ikx) \tag{9.1}$$

where the momentum eigenvalue is $\hbar k$. Wave functions corresponding to different values of k should be orthogonal and the constants A_k should be chosen to ensure that the wave functions are normalized. However, functions of the form (9.1) extend throughout the one-dimensional space and the usual normalization procedure would imply that $A = 0$, corresponding to the fact that, if the particle can be anywhere between plus and minus infinity with equal probability, then the probability of finding it in the vicinity of any particular point must be zero. We can overcome this problem by redefining ψ_k so as to represent a *beam* containing many particles rather than a single particle, and we assume that the average separation of the particles in the beam is L. If we now normalize ψ_k so that $|\psi_k(x)|^2\,dx$ represents the probability of finding *any* particle in the region between x and $x + dx$ we get

$$\int_0^L \psi_k^* \psi_k \, dx = 1 \tag{9.2}$$

and hence, substituting from (9.1)

$$\psi_k = L^{-1/2} \exp(ikx) \tag{9.3}$$

It would clearly be very convenient if the orthogonality condition could be expressed in a similar manner so that

$$\int_0^L \psi_k^* \psi_{k'} \, dx = \delta_{kk'} \tag{9.4}$$

Substitution of (9.3) into the left-hand side of (9.4) shows that the latter equation can be valid only if the values of k are restricted so that

$$k = 2n\pi/L \tag{9.5}$$

where n is an integer. This condition is identical to that derived in our discussion of the one-dimensional solid in Chapter 7 where we imposed 'periodic boundary conditions' so that $\psi(x + L) = \psi(x)$ and we shall assume these to be valid throughout our ensuing discussions. No physical significance should be attached to the periodic boundary conditions: they are a mathematical device which enables us to impose orthonormality on free particle wave functions and hence to calculate properties such as the density of states (see below). We note that it is not necessary that L be defined as the mean particle separation for periodic boundary conditions to be applicable, and indeed in Chapter 7 this was not the case. However, when discussing scattering experiments we shall assume that L represents the average separation of the particles in the incident beam and that the amplitude of the wave function representing scattered particles is related to this via the scattering probability.

Particle Flux

A property of particle beams that will be particularly useful in the subsequent discussion is known as particle flux. In one dimension this is defined as the average number of particles passing a point per unit time and is represented by S. Clearly if the particles are in a momentum eigenstate, S equals the number of particles per unit length multiplied by the particle velocity: that is,

$$S = \hbar k/mL \tag{9.6}$$

where m is the particle mass. An expression for the flux in the general case where the one-dimensional wave function is $\Psi(x, t)$ is derived below.

Consider a region of the x axis between the points $x = x_1$ and $x = x_2$ where $x_2 > x_1$. The probability of finding a particle in this region is therefore P where

$$P = \int_{x_1}^{x_2} \psi^* \psi \, dx \tag{9.7}$$

The net flux of particles into this region must be equal to the rate of change in time of P, otherwise particles would have to appear or disappear inside the region so that the total number of particles would not be conserved and, although such creation and destruction of particles can occur (particularly in the case of photons, for example), we shall not include any such cases in our discussion. It follows then that

$$S(x_1) - S(x_2) = \frac{\partial P}{\partial t} = \int_{x_1}^{x_2} \left(\Psi^* \frac{\partial \Psi}{\partial t} + \Psi \frac{\partial \Psi^*}{\partial t} \right) dx \tag{9.8}$$

We can now substitute for $\partial \Psi/\partial t$ and $\partial \Psi^*/\partial t$ from the time-dependent Schrödinger equation (8.1) to get

$$S(x_1) - S(x_2) = \frac{i\hbar}{2m} \int_{x_1}^{x_2} \left(\Psi^* \frac{\partial^2 \Psi}{\partial x^2} - \Psi \frac{\partial^2 \Psi^*}{\partial x^2} \right) dx \tag{9.9}$$

where terms involving the potential energy have cancelled each other out. We now integrate by parts to obtain

$$S(x_1) - S(x_2) = \frac{i\hbar}{2m}\left[\Psi^* \frac{\partial \Psi}{\partial x} - \Psi \frac{\partial \Psi^*}{\partial x} \right]_{x_1}^{x_2} \tag{9.10}$$

Equation (9.10) is valid for all pairs of points x_1 and x_2, so the flux past the point x must be given by

$$S(x) = -\frac{i\hbar}{2m}\left(\Psi^* \frac{\partial \Psi}{\partial x} - \Psi \frac{\partial \Psi^*}{\partial x} \right) \tag{9.11}$$

We note that, if we substitute the momentum eigenfunction (9.3) into (9.11), we get the same expression (9.6) for the particle flux as was obtained earlier.

Scattering by a Potential Step

We now consider the case where a beam of particles is scattered by a simple potential step of the form illustrated in Fig. 9.1. That is, we are considering the motion of particles in a potential $V(x)$ where

$$\left.\begin{array}{ll} V(x) = 0 & x \leqslant 0 \\ V(x) = V_0 & x > 0 \end{array}\right\} \tag{9.12}$$

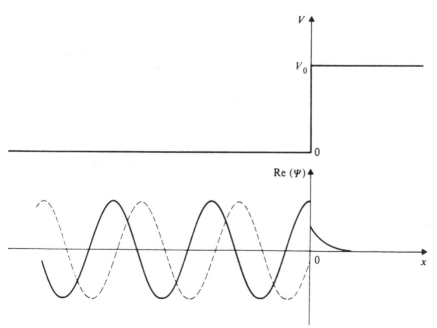

Figure 9.1 The potential V and the real parts of the incident, reflected, and transmitted waves in the case of the scattering of particles by a potential step.

We are interested in the steady-state solution to the problem, ignoring any transient effects associated with the 'switching on' of the beam, which means that, as the potential (9.12) is time independent, the wave function should be one of the energy eigenfunctions of the system. Now in the region where the potential is zero the momentum eigenfunctions with eigenvalues $\pm\hbar k$ are also degenerate eigenfunctions of the energy with energy $\hbar^2 k^2 / 2m$. A general expression for the energy eigenfunction in the region $x < 0$ is therefore

$$u(x) = L^{-1/2}(e^{ikx} + \alpha e^{-ikx}) \qquad (9.13)$$

where α is a constant. In studying scattering we wish to predict the results of experiments that measure the flux of the incident and scattered wave functions separately. In one dimension this means that we have one detector in the region $x < 0$ which identifies and counts particles moving in (say) the positive direction while leaving the others undisturbed, and another which performs a similar measurement on the particles moving in the negative direction. The energy eigenfunction (9.13) then represents the wave function *before* these measurements are performed and the probable results can then be predicted using the standard procedure set out in the basic postulates. Thus, in the present case (9.13) tells us that a measurement of the flux of particles moving in the positive k direction will produce a result $\hbar k / mL$ while that in the negative direction is $|\alpha|^2 \hbar k / mL$, so that the probability of a particular particle being scattered is $|\alpha|^2$.

To obtain an expression for the constant α we must consider the wave function in the region $x > 0$ where its form depends on whether the incident kinetic energy $E = \hbar^2 k^2 / 2m$ is greater or less than V_0. We shall consider the latter case first when the solution to the time-independent Schrödinger equation is easily seen to be

$$u(x) = C \exp(-\kappa x) \qquad x > 0 \qquad (9.14)$$

where $\kappa = [2m(V_0 - E)/\hbar^2]^{1/2}$ and we have rejected the corresponding solution with a positive exponent because it implies a divergence of u as x tends to infinity, in breach of the boundary conditions applying to one-dimensional systems.

We can now apply the conditions that the wave function and its first spatial derivatives be continuous at $x = 0$ (cf. Chapter 2) to get

$$L^{-1/2}(1 + \alpha) = C$$
$$L^{-1/2}ik(1 - \alpha) = -\kappa C$$

and hence

$$\left.\begin{array}{l} \alpha = -(\kappa + ik)/(\kappa - ik) \\ C = -[2ik/(\kappa - ik)]L^{-1/2} \end{array}\right\} \qquad (9.15)$$

The scattering probability $|\alpha|^2$ is therefore seen to be unity so that all the particles incident on the potential step are scattered and none is transmitted.

This point can be confirmed by substituting from (9.14) into (9.11) which produces a zero value for the transmitted flux. The net result of the scattering is therefore to produce a scattered wave whose amplitude is the same as that of the incident wave and whose phase at the point $x = 0$ is increased relative to the phase of the incident wave by an amount $\delta = \tan^{-1}[2k\kappa/(\kappa^2 - k^2)]$. Clearly a similar result to this will apply in any one-dimensional scattering problem where the potential barrier is high enough to prevent the particles being transmitted, even if it does not have the simple step form discussed above. The total wave function in a region on the incident side of the scatterer where the potential is zero will always consist of a sum of incident and reflected plane waves of equal amplitude, and the detailed nature of the scattering potential will affect only the phase shift δ. Similar phase shifts turn out to be of considerable importance in three-dimensional scattering and we shall return to this concept later in the chapter when we discuss the method of 'partial waves'.

We now turn to the case where the kinetic energy of the particles is greater than the height of the barrier so that we expect there to be a finite probability of the particle being transmitted. The wave function in the region $x < 0$ still has the general form (9.12) while the transmitted wave is clearly

$$u(x) = C' \exp(ik'x) \tag{9.16}$$

where

$$k' = [2m(E - V_0)/\hbar^2]^{1/2}$$

and C' is obtained along with α by applying the continuity conditions

$$\left. \begin{aligned} \alpha &= \frac{k - k'}{k + k'} \\[2mm] C' &= \frac{2k}{k + k'} L^{-1/2} \end{aligned} \right\} \tag{9.17}$$

The reflected flux is therefore

$$\frac{\hbar k}{mL}|\alpha|^2 = \frac{\hbar k}{mL}\frac{(k - k')^2}{(k + k')^2}$$

while the transmitted flux is

$$\frac{\hbar k'}{m}|C'|^2 = \frac{\hbar k}{mL}\frac{4kk'}{(k + k')^2}$$

It follows directly that the scattering and transmission probabilities are

$$\frac{(k - k')^2}{(k + k')^2} \quad \text{and} \quad \frac{4kk'}{(k + k')^2}$$

respectively and we note that the sum of these two probabilities is equal to one as expected.

9.2 SCATTERING IN THREE DIMENSIONS

We now turn to the three-dimensional case and begin with some general statements about particle flux and scattering probability. Whereas in one dimension the particle flux was defined as the number of particles passing a point per second, in three dimensions we define *flux density* as a vector quantity **S** such that $\mathbf{S} \cdot d\mathbf{A}$ is the total flux of particles passing through the element of area $d\mathbf{A}$ per second. Clearly the direction of **S** corresponds to the direction of motion of the particles at the point under consideration, while its magnitude represents the number of particles crossing unit area per second. We shall now obtain an expression for **S** in the case of a system represented by the wave function $\Psi(\mathbf{r}, t)$ using a procedure similar to that employed in the one-dimensional case.

Consider a volume V enclosed by a surface A. The net number of particles entering V through A in unit time must be equal to the rate of increase of the probability of finding a particle in V. That is

$$-\int_A \mathbf{S}(\mathbf{r}) \cdot d\mathbf{A} = \frac{\partial}{\partial t} \int_V \Psi^* \Psi \, d\tau$$

$$= \int_V \left(\frac{\partial \Psi^*}{\partial t} \Psi + \Psi^* \frac{\partial \Psi}{\partial t} \right) d\tau \tag{9.18}$$

Following a procedure similar to that applied in the one-dimensional case—(9.9) to (9.11)—we substitute for $(\partial \Psi / \partial t)$ and $(\partial \Psi^* / \partial t)$ from the time-dependent Schrödinger equation, the potential energy terms cancel out as before and we get

$$\int_A \mathbf{S}(\mathbf{r}) \cdot d\mathbf{A} = -\frac{i\hbar}{2m} \int (\Psi^* \nabla^2 \Psi - \Psi \nabla^2 \Psi^*) \, d\tau$$

$$= -\frac{i\hbar}{2m} \int_A (\Psi^* \nabla \Psi - \Psi \nabla \Psi^*) \cdot d\mathbf{A} \tag{9.19}$$

where we have applied the theorem in vector calculus known as Green's theorem which corresponds to integration by parts in three dimensions. The expression for $\mathbf{S}(\mathbf{r})$ follows directly, remembering that (9.19) is valid for any closed surface:

$$\mathbf{S}(\mathbf{r}) = -\frac{i\hbar}{2m} (\Psi^* \nabla \Psi - \Psi \nabla \Psi^*) \tag{9.20}$$

which is therefore the three-dimensional equivalent of (9.11).

As an example we consider the special case of a beam of free particles of momentum $\hbar\mathbf{k}$ whose wave function is

$$\Psi(\mathbf{r}, t) = V^{-1/2} \exp i(\mathbf{k} \cdot \mathbf{r} - Et/\hbar) \tag{9.21}$$

where $E = \hbar^2 k^2/2m$ and the beam has been normalized so that it contains on average one particle in a volume V (which is of course not necessarily the same as the volume V used above). Substituting from (9.21) into (9.20) we have

$$\mathbf{S} = \frac{\hbar \mathbf{k}}{mV} \tag{9.22}$$

which is the same expression as would be obtained from elementary considerations remembering that in this case all particles have velocity $\hbar k/m$.

Scattering Cross Section

If a beam of particles is incident on a scattering object some particles will be scattered while others will pass on undisturbed, and we can therefore define a probability that scattering will take place. Clearly the probability of scattering will be proportional to the flux density of the incident beam, and the probability that a particle will be scattered from a beam of unit flux density—that is, where one particle passes through unit area per second—in unit time is known as the scattering cross section, σ. It follows that σ has the dimensions of area and, in the classical case of a beam of small particles which interact with a scattering object only when they strike it, it is equal to the geometrical cross section of the body in a plane perpendicular to the beam. In quantum mechanics, however, such a simple interpretation is rarely possible.

We are often interested, not only in the total probability of scattering, but also in the probability that the particles are scattered in a particular direction, which is usually defined by the spherical polar angles θ and ϕ referred to the direction of the incident beam as the polar axis. To this end we define the *differential cross section* $\sigma(\theta, \phi)$ such that, for an incident beam of unit flux density, $\sigma(\theta, \phi)\,d\Omega$ is the probability per second of a particle being scattered into the element of solid angle $d\Omega$ around the direction defined by θ and ϕ. If we express $d\Omega$ in terms of θ and ϕ in the usual way

$$d\Omega = \sin\theta\, d\theta\, d\phi \tag{9.23}$$

it follows directly from the definitions of the total and differential cross sections that

$$\sigma = \int \sigma(\theta, \phi)\, d\Omega$$
$$= \int\int \sigma(\theta, \phi) \sin\theta\, d\theta\, d\phi \tag{9.24}$$

In many cases the problem has cylindrical symmetry about an axis parallel to the incident beam and the differential cross section is consequently independent of ϕ and written as $\sigma(\theta)$. In this case the probability per second of scattering into the element of angle $d\theta$ around θ, irrespective of the value of ϕ, is clearly given by $2\pi\sigma(\theta) \sin\theta\, d\theta$.

Centre of Mass Frame

Finally in this section we note that the scattering experiments are often carried out using beams of particles whose mass is comparable to the mass of the particles in the target (e.g., the scattering of neutrons by the protons in hydrogen referred to near the end of this chapter). As is shown in Chapter 10, such a situation can be treated by considering the corresponding case of a particle of mass μ equal to the reduced mass of the two particles—that is, $\mu = m_1 m_2/(m_1 + m_2)$—interacting with a fixed target. We can therefore apply all the theory relating to the scattering of particles from fixed objects to this case and obtain results referred to a frame of reference attached to the centre of mass of the system which may then be transferred back into the laboratory frame by standard methods.

9.3 THE BORN APPROXIMATION

We saw in the earlier discussion of one-dimensional scattering that the energy eigenfunction of the system could be expressed as linear combinations of incident and scattered waves, and the scattering probability could then be calculated from the expansion coefficients. A similar procedure can in principle be followed in the three-dimensional case, but the initial solution of the time-independent Schrödinger equation for the energy eigenfunctions is generally very difficult and often impossible. In the next section we shall discuss a method for doing this in the particular case of a spherically symmetric scattering potential, but for the moment we shall describe a procedure known as the *Born approximation*. This method is based on first-order perturbation theory and is usually valid when the average energy of the interaction between an incident particle and the scatterer is much smaller than the particle's kinetic energy.

We consider the case of a beam of particles approaching a scattering object along a direction parallel to the vector k_0. Except in the vicinity of the scatterer, the potential is zero and the incident beam can therefore be described by a plane wave function whose time-independent part is u_0 where

$$u_0 = V^{-1/2} \exp(i k_0 \cdot r) \tag{9.25}$$

We have assumed this wave function to be normalized so that there is on average one particle in the volume V, and we also impose periodic boundary conditions, similar to those applied in the one-dimensional case, so that the allowed values of the Cartesian components of k_0 are

$$k_{0x} = 2n_1 \pi/L_1 \qquad k_{0y} = 2n_2 \pi/L_2 \qquad k_{0z} = 2n_3 \pi/L_3 \tag{9.26}$$

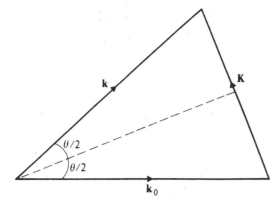

Figure 9.2 The relationship between the scattering angle (θ), the scattering vector ($\mathbf{K}$), and the wave vectors of the incident ($\mathbf{k}_0$) and scattered ($\mathbf{k}$) waves.

where n_1, n_2, and n_3 are integers and $L_1 L_2 L_3 = V$. We now imagine that scattering has taken place and that a scattered particle has been detected moving in some direction $\mathbf{k}$ (cf. Fig. 9.2). The normalized wave function will now be u_1 where

$$u_1 = V^{-1/2} \exp(i\mathbf{k} \cdot \mathbf{r}) \tag{9.27}$$

and boundary conditions similar to (9.26) are imposed on the Cartesian components of $\mathbf{k}$. Remembering that the scattering potential is to be treated as a perturbation, we see that u_0 and u_1 are both energy eigenfunctions of the unperturbed (zero-potential everywhere) problem. We can therefore represent the scattering process as a *transition* from the state u_0 to the state u_1 and relate the scattering probability and hence the scattering cross section to the transition probability.

We previously calculated an expression for the transition probability when we considered time-dependent perturbation theory in Chapter 8. Although the potential is not time-dependent in the present case, an identical argument can be applied (after putting $\omega = 0$) as far as Eq. (8.29). At this point we see that transitions occur only if the energies of the initial and final states are identical (that is, $\omega_{mn} = 0$, if $|\mathbf{k}_0| = |\mathbf{k}|$ and the scattering is elastic) when *both* of the terms on the right-hand side of (8.29) are simultaneously non-zero. This introduces a factor of two into the subsequent equations and a consequent factor of four into the Fermi-golden-rule expression (8.34) for the transition rate W which is therefore

$$W = \frac{2\pi}{\hbar^2} |U_{\mathbf{k}\mathbf{k}_0}|^2 g(\omega) \tag{9.28}$$

The matrix element $U_{\mathbf{k}\mathbf{k}_0}$ is calculated using the approximate eigenfunctions u_0 and u_1 and the scattering potential $U(\mathbf{r})$ (note that we now use U to represent the potential in order to avoid confusion with the volume V):

$$U_{\mathbf{k}\mathbf{k}_0} = \int u_1^* U(\mathbf{r}) u_0 \, d\tau$$

$$= \frac{1}{V} \int U(\mathbf{r}) \exp(-i\mathbf{K}\cdot\mathbf{r}) \, d\tau \tag{9.29}$$

where $\mathbf{K} = \mathbf{k} - \mathbf{k}_0$ and is known as the *scattering vector*.

To calculate the differential cross section using (9.28) we must obtain an expression for the number of states in the angular frequency range ω to $\omega + d\omega$ available to a particle scattered into the element of solid angle $d\Omega$. It follows directly from the restrictions on the allowed values of $\mathbf{k}$ (9.26) that the number of states in an element of volume $d^3\mathbf{k} = dk_x \, dk_y \, dk_z$ in k-space is equal to

$$(L_1 L_2 L_3 / 8\pi^3) d^3\mathbf{k} = (V/8\pi^3) \, d^3\mathbf{k} \tag{9.30}$$

where we have assumed that V is large enough for the volume of the element $d^3\mathbf{k}$ to be large compared with that of the k-space cell, $8\pi^3/V$. If we now consider the states whose wave vectors lie in the element of solid angle $d\Omega$ around the direction $\mathbf{k}$ and whose magnitudes lie between k and $k + dk$, these occupy a k-space volume equal to $k^2 \, d\Omega \, dk$ and the total number of such states is therefore

$$(V/8\pi^3) k^2 \, d\Omega \, dk \tag{9.31}$$

But we know that the energy of a free particle is $E = \hbar^2 k^2/2m$ so that

$$dE = (\hbar^2 k/m) \, dk$$

and therefore $d\omega = (\hbar k/m) \, dk$ where $E = \hbar\omega$. Writing the required density of states as dg to emphasize the fact that it refers to an element of solid angle $d\Omega$, the number of states with angular frequency between ω and $\omega + d\omega$, and whose k-vectors lie within $d\Omega$, equals $dg \, d\omega$ and we have

$$dg = \frac{mkV}{8\pi\hbar} \, d\Omega \tag{9.32}$$

We can now obtain an expression for the probability per unit time (dW) of a particle being scattered from a state u_0 to a state whose wave vector lies within the element of solid angle $d\Omega$ around the direction $\mathbf{k}$, by substituting the expression (9.32) for dg in place of g in (9.28) and using (9.29)

$$dW = \frac{mk}{4\pi^2 V\hbar^3} \left| \int U(\mathbf{r}) e^{-i\mathbf{K}\cdot\mathbf{r}} \, d\tau \right|^2 d\Omega \tag{9.33}$$

Finally, to calculate the differential cross section we must divide the right-hand side of (9.33) by the magnitude of the incident flux $(\hbar k/mV)$

and by $d\Omega$ to give

$$\sigma(\theta, \phi) = \frac{m^2}{4\pi^2\hbar^4} |\int U(\mathbf{r}) e^{-i\mathbf{K}\cdot\mathbf{r}} d\tau|^2 \qquad (9.34)$$

Thus, provided we know the form of the scattering potential, the scattering cross section can be calculated by evaluating the Fourier transform of $U(\mathbf{r})$ and hence the right-hand side of (9.34). The above result is equivalent to that which would be obtained on the basis of Fraunhofer diffraction theory used in optics, where the amplitude of the scattered light as a function of scattering vector is proportional to the Fourier transform of the diffracting object.

We shall shortly discuss a couple of examples of the application of the Born approximation when the scattering potential is spherically symmetric, in which case the volume integral in (9.34) can be partly evaluated using spherical polar coordinates r, θ', and ϕ' referred to the direction of $\mathbf{K}$ as polar axis. (N.B. θ' and ϕ' should be distinguished from θ and ϕ which define the direction of $\mathbf{k}$ relative to k_0) Thus

$$\int U(\mathbf{r}) e^{-i\mathbf{K}\cdot\mathbf{r}} d\tau = \int_0^{2\pi} \int_0^\pi \int_0^\infty U(r) e^{-iKr\cos\theta'} r^2 \, dr \sin\theta' \, d\theta' \, d\phi'$$

$$= \frac{4\pi}{K} \int_0^\infty U(r) r \sin kr \, dr \qquad (9.35)$$

and the differential cross section becomes

$$\sigma(\theta) = \frac{4m^2}{\hbar^4 K^2} \left| \int_0^\infty U(r) \sin Kr \, dr \right|^2 \qquad (9.36)$$

We note that the differential cross section is independent of ϕ and that its dependence on θ is through the magnitude of the scattering vector K which is seen from Fig. 9.2 to be equal to $2k_0 \sin(\theta/2)$.

Example 9.1 Scattering by a spherical potential well or step Our first example relates to scattering by a spherically symmetric potential given by

$$\begin{aligned} U(r) &= 0 \qquad r > a \\ &= U_0 \qquad r \leqslant a \end{aligned} \right\} \qquad (9.37)$$

where U_0 and a are constants. U_0 may be negative (potential well) or positive (potential step) but in any case is to be assumed small enough for the Born approximation to be applied. The more general case including larger U_0 will be treated by the method of partial waves towards the end of this chapter. The potential is spherically symmetric so (9.36) can be used and the integral on the right-hand side of this equation is

$$U_0 \int_0^a r \sin Kr \, dr = U_0 (\sin Ka - Ka \cos Ka)/K^2 \qquad (9.38)$$

This can be substituted into (9.36) to produce a rather complicated expression for the scattering cross section.

In the particular case where Ka is much less than one, the trigonometric terms can be expanded as series of ascending powers of Ka and the first non-vanishing term in (9.38) is equal to $\frac{1}{3}U_0 Ka^3$. The differential cross section in this limit is then

$$\sigma(\theta) = \frac{4m^2 U_0^2 a^6}{9\hbar^4} \qquad (9.39)$$

Thus, if the geometrical radius of the scatterer is much less than the wavelength ($\lambda = 2\pi/k$) associated with the incident particles, Ka will be small for all scattering angles, the scattering will be isotropic (i.e., independent of θ), and the total cross section will be simply 4π times that given in (9.39). The fact that objects whose dimensions are much less than λ scatter isotropically also follows more generally from Eq. (9.34) where we see that if $\mathbf{K} \cdot \mathbf{r}$ is much less than one for all values of $\mathbf{r}$ for which U is neither zero nor insignificantly small, then the differential cross section is independent of angle. It is also in agreement with the familiar phenomenon in optics in which light striking an object whose dimensions are much smaller than the wavelength is scattered equally in all directions.

Example 9.2 Rutherford scattering We now consider scattering by a potential whose magnitude is inversely proportional to the distance from the origins of the particle being scattered. That is,

$$U(r) = \beta r^{-1} \qquad (9.40)$$

where β is a constant. This is known as Rutherford scattering because it can be applied to the case of positively charged α particles approaching a target containing atoms with positively charged nuclei (when $\beta = 2Ze^2/4\pi\varepsilon_0$), and this arrangement was first studied by Rutherford in 1911.

Substituting from (9.40) the integral in (9.36) becomes

$$\beta \int_0^\infty \sin Kr \, dr = -\frac{\beta}{K} [\cos Kr]_0^\infty \qquad (9.41)$$

The integral is indeterminate at its upper limit, but in practice the Coulomb potential is always modified at large distances (e.g., due to screening by the atomic electrons) so that at large r it falls off more rapidly than is implied by (9.40); the integral is then negligibly small at large r and we obtain the following expression for the differential cross section:

$$\sigma(\theta) = \frac{4m^2\beta^2}{\hbar^4 K^4}$$

$$= \frac{m^2\beta^2}{4\hbar^4 k_0^4 \sin^4(\theta/2)}$$

$$= \frac{m^2\beta^2}{4p_0^4 \sin^4(\theta/2)} \qquad (9.42)$$

where p_0 is the magnitude of the momentum of the incident particles and θ is the angle between the incident and scattered directions as before. It was Rutherford's confirmation of this formula by the observation of the scattering of α particles from gold foil that led to his postulate of the nuclear atom; of course this development preceded quantum mechanics, but by a happy coincidence it turns out that the classical and quantum expressions are identical in this case.

9.4 PARTIAL WAVE ANALYSIS

In the final section of this chapter we consider an approach to the scattering problem which is in some ways the opposite to that adopted in the Born approximation. Whereas the latter expresses the cross section in terms of the Fourier transform of the scattering potential—that is, the potential is expanded as a linear combination of plane waves whose form matches that of the wave functions of the incident and scattered particles—in partial wave analysis we start with the eigenfunctions of the scattering potential and express the incident plane wave as a linear combination of these. The method is particularly useful if the scattering potential is spherically symmetric, and this is the only case we shall consider. Unlike the Born approximation, no limitations are placed on the strength of the interaction between the particle beam and the scatterer.

The principle of the method is similar to that adopted for the one-dimensional example discussed earlier. We first obtain the wave function as an energy eigenfunction whose eigenvalue equals the energy of the incident particles. This is then expressed as a sum of a plane wave representing the incident beam and a scattered wave, and we use the measurement postulate to calculate the differential cross section as the ratio of the flux of the scattered wave through an element of solid angle in a particular direction, to the flux density of the incident wave. For this procedure to be applicable, the

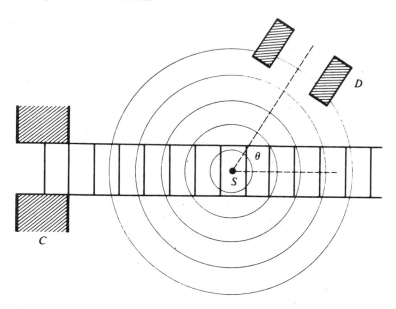

Figure 9.3 Particles passing through the collimator C are represented by a plane wave which is scattered by the scatterer S. Only the scattered wave enters the detector D, but there is a region around the scatterer, much larger than the scatterer itself, over which the wave function is a linear combination of the incident and scattered waves.

experimental measurement must distinguish between the incident and scattered waves and a typical set-up to achieve this is shown in Fig. 9.3. The incident beam is defined by a collimator whose diameter is small compared with the distance from the scatterer to the detector, so ensuring that particles cannot enter the detector without first being scattered, and that the wave function in the region of the detector is therefore just that of the scattered wave. On the other hand, the collimator diameter must be much larger than both the wavelength of the incident beam and the dimensions of the scattering object. This ensures both that the incident particles can be represented by a plane wave, and also that there is a region around the scatterer, much larger than the scatterer itself, over which the wave function is a linear combination of the incident plane wave and the scattered wave.

We begin our discussion by considering the form of the energy eigenfunctions associated with a spherically symmetric potential $U(r)$. This problem was discussed in Chapter 3, but at that point we restricted our attention to bound states of potential wells where the total energy was found to be quantized. In the scattering case, however, we are interested in wave functions representing particles approaching the scatterer from a large distance and leaving it again. Thus, either the potential is such that binding is impossible (e.g., if it is repulsive at all distances) or, if bound states do exist, the incident energy is too great for binding to occur. This means that the energy levels of the scattering system are not confined to a discrete set of values.

To obtain the energy eigenfunctions of eigenvalue E for a spherically symmetric potential $U(r)$, we must solve the time-independent Schrödinger equation:

$$-\frac{\hbar^2}{2m} \nabla^2 u + U(r)u = Eu \tag{9.43}$$

This equation can be separated in spherical polar coordinates just as was done for a bound particle in Chapter 3, by putting

$$u = R(r)Y_{lm}(\theta, \phi) \tag{9.44}$$

where the Y_{lm} are spherical harmonics which are defined in that chapter and R is obtained by solving the radial equation

$$-\frac{\hbar^2}{2m}\frac{1}{r^2}\frac{d}{dr}\left(r^2 \frac{\partial R}{\partial r}\right) + \frac{l(l+1)\hbar^2}{2mr^2} R + UR = ER \tag{9.45}$$

If we now define k and $\chi(r)$ so that $R = \chi/r$ and $E = \hbar^2 k^2 / 2m$, (9.45) becomes

$$\frac{d^2\chi}{dr^2} + \left(k^2 - \frac{l(l+1)}{r^2} - \frac{2mU}{\hbar^2}\right)\chi = 0 \tag{9.46}$$

In a scattering experiment we are interested in the wave function at a large distance from the scattering object. We can therefore obtain a lot of information from the asymptotic form of the solutions at large r when we can neglect the last two terms in the bracket—assuming that the potential $U(r)$ is effectively zero at large r, which is nearly always the case. Equation (9.46) then becomes

$$\frac{d^2\chi}{dr^2} + k^2\chi = 0 \tag{9.47}$$

the general solution to which can be written as

$$\chi = A\, e^{ikr} + B\, e^{-ikr} \tag{9.48}$$

where A and B are constants. Using (9.44) we obtain the following expressions for the asymptotic form of the energy eigenfunctions which we now write as u_{klm}:

$$u_{klm} = \frac{1}{r}(A\, e^{ikr} + B\, e^{-ikr})Y_{lm}(\theta, \phi) \tag{9.49}$$

The energy eigenvalues are independent of the quantum numbers l and m, so any linear combination of the functions u_{klm} that have the same value of k is also an eigenfunction. We shall shortly obtain such a linear combination with coefficients chosen so that it has the desired form of a sum of the incident plane wave and a scattered wave, but in the meantime useful information can be obtained by considering the case where the wave function is just one of the eigenfunctions (9.49). The first term in this expression represents a wave travelling radially outwards while the second corresponds to a wave travelling inwards. This wave function would therefore apply to an experiment where particles approached the scatterer along the radial direction with the numbers coming in at different angles (θ, ϕ) proportional to $|Y_{l,m}|^2$. The spherical harmonics, $Y_{l,m}$ are illustrated in Fig. 3.3, and we see, for example, that if $l = m = 0$ all directions of approach are equally probable, while if $l = 1$ and $m = 0$, the most favoured direction of approach is along the z axis. For such an experiment, equation (9.49) tells us that the angular part of the wave function after the scattering will be identical to that beforehand. The total number of particles (N_1) passing outwards per second through the surface of a sphere of radius r centred on the origin is readily obtained by substituting the first term in (9.49) (which we refer to below as u_1) into the general expression for the flux (9.20) and integrating over all solid angles, remembering that the radial component of $\nabla \psi$ is $\partial \psi / \partial r$. Thus

$$
\begin{aligned}
N_1 &= -\frac{i\hbar}{2m} \int_0^{2\pi} \int_0^{\pi} \left(u_1^* \frac{\partial u_1}{\partial r} - u_1 \frac{\partial u_1^*}{\partial r} \right) r^2 \sin \theta \, d\theta \, d\phi \\
&= -\frac{i\hbar}{2m} \left(2|A|^2 \frac{ik}{r^2} \right) r^2 \int_0^{2\pi} \int_0^{\pi} |Y_{lm}|^2 \sin \theta \, d\theta \, d\phi \\
&= \frac{\hbar k}{m} |A|^2
\end{aligned}
\tag{9.50}
$$

remembering that the spherical harmonics are normalized. We note that N_1 is independent of r as it must be because no particles are lost or gained as the distance from the scatterer is varied. The number flowing inward per second, N_2, is similarly calculated using the second term in (9.49) as

$$
N_2 = \frac{\hbar k}{m} |B|^2
\tag{9.51}
$$

We now note that in a spherically symmetric system the energy eigenfunctions (9.49) are also eigenfunctions of the total angular momentum and its z component so that these two quantities must be conserved in an energy eigenstate. Thus no change in the quantum numbers l and m can occur as a result of the scattering process and all particles incident on the scatterer in a state described by the second term in (9.49) must leave in a state described by the first term. It follows that N_1 and N_2 must be equal so that the constants A and B can differ only by a phase factor. We can therefore write

$$
u_{klm} = \frac{A}{r} \left[e^{-ikr} - e^{i(kr - l\pi + 2\delta_{lm})} \right] Y_{lm}(\theta, \phi)
\tag{9.52}
$$

where we have defined the *phase shift* δ_{lm} so that the phase factor relating B to A is $-\exp i(2\delta_{lm} - l\pi)$ as this form is particularly convenient for the later discussion. We conclude that, if an experiment were performed in which the incident particles have a wave function of the form

$$\frac{A}{r} e^{-ikr} Y_{lm}(\theta, \phi) \tag{9.53}$$

then all the particles would be scattered and the form of the scattered wave would be identical with that of the incident wave apart from a change of direction and a constant phase factor. The actual value of the phase shift in any particular case is determined by the form of the scattering potential. We note the similarity between this result and that obtained in our earlier example of scattering by a potential step in one dimension (cf. 9.15).

In practice, of course, scattering experiments are not performed with incident waves of the form (9.53) which represent particles approaching the scatterer from all sides with an angular distribution determined by the spherical harmonic. Instead, as we have seen earlier, beams of particles are used all of which are travelling in the same (say the z) direction so that the incident wave function u_{k0} has the plane wave form

$$u_{k0} = V^{-1/2} \exp(ikz)$$
$$= V^{-1/2} \exp(ikr \cos\theta) \tag{9.54}$$

in spherical polar coordinates. In order to proceed further we have to express the plane wave (9.54) as a linear combination of incoming and outgoing spherical waves of the form discussed above. This is a standard mathematical expression (see, for example, G. N. Watson, *A Treatise on the Theory of Bessel Functions*, Cambridge University Press, 1958) which in the asymptotic limit of large r, has the form

$$V^{-1/2} e^{ikr\cos\theta} = \tfrac{1}{2} V^{-1/2} \sum_{l=0}^{\infty} (2l+1) i^{2l+1} \left[\frac{1}{kr} e^{-ikr} - \frac{1}{kr} e^{i(kr - l\pi)} \right] P_l(\cos\theta) \tag{9.55}$$

where $P_l(\cos\theta)$ is the Legendre polynomial of order l and we remember from Chapter 3 that this is related to the spherical harmonic Y_{l0} by

$$Y_{l0} = \left(\frac{2l+1}{4\pi} \right)^{1/2} P_l(\cos\theta) \tag{9.56}$$

If we compare (9.55) and (9.56) with (9.52) we see that a plane wave travelling in the positive z direction can be expressed as a linear combination of incoming and outgoing spherical waves—known as 'partial waves'—which all have $m = 0$ and different values of l: the phase relation between the incoming and outgoing partial waves in this case is such that the phase shifts δ_{lm} in (9.52) are all equal to zero.

Given that (9.55) is a representation of the incident plane wave we can now consider how this is modified by the presence of the scattering object. We remember that in the steady state the total wave function must be an energy eigenfunction and must therefore be equal to a linear combination of the eigenfunctions determined earlier (9.52) that have the same value of k as the incident wave. Thus

$$\psi = \sum_{lm} \frac{A_{lm}}{r} [e^{-ikr} - e^{i(kr - l\pi + 2\delta_{lm})}] Y_{lm}(\theta, \phi) \tag{9.57}$$

On the other hand the total wave function can also be written as a sum of an incident and a scattered wave and we know that when these are separated in the course of the experiment (see Fig. 9.3) the latter represents only particles which are moving outwards from the scatterer. The scattered wave therefore must not contain any components with negative values of the exponent ikr and the coefficients of such terms in the total wave function (9.57) must be the same as in the incident wave (9.55). Thus we have, using (9.56),

$$A_{lm} = 0 \qquad m \neq 0$$

$$A_{l0} = \tfrac{1}{2}[4\pi(2l + 1)/V]^{1/2} i^{2l+1} k^{-1}$$

The first of these conditions proves that the wave function is independent of ϕ, as would be expected from the symmetry of the problem. If we now substitute back into (9.57) and make use of (9.56) again we get

$$\psi = \tfrac{1}{2} V^{-1/2} \sum_{l=0}^{\infty} (2l + 1) i^{2l+1} \left[\frac{1}{kr} e^{-ikr} - \frac{1}{kr} e^{i(kr + 2\delta_l - l\pi)} \right] P_l(\cos \theta) \tag{9.58}$$

where we have rewritten δ_{lm} as δ_l, because (9.58) contains only terms with $m = 0$. We now subtract the incident wave (9.55) from the total wave function (9.58) to get an expression for the scattered wave function ψ_s which then has the form

$$\psi_s = V^{-1/2} \frac{1}{r} e^{ikr} f(\theta) \tag{9.59}$$

where

$$f(\theta) = \frac{1}{2k} \sum_{l=0}^{\infty} (2l + 1) i^{2l+1} e^{-il\pi} (1 - e^{2i\delta_l}) P_l(\cos \theta)$$

$$= \frac{1}{k} \sum_{l=0}^{\infty} (2l + 1) e^{i\delta_l} \sin \delta_l P_l(\cos \theta) \tag{9.60}$$

We note that if all the phase shifts δ_l are zero, $f(\theta)$ vanishes and there is no scattering, which is consistent with the fact that the total wave function (9.58) is then identical with the incident plane wave (9.55). The extent and nature of the scattering can therefore be calculated, given the plane shifts δ_l, and we shall shortly discuss how to evaluate these in particular cases. First, however, we use the results derived so far to obtain expressions for the differential and total scattering cross sections.

To evaluate the differential cross section we must calculate the number (dN) of scattered particles crossing an area $r^2\,d\Omega$ on the surface of a sphere of radius r per second. Using the general expression for the flux (9.19) and remembering that the radial component of $\nabla\psi_s$ is $\partial\psi_s/\partial r$ we get

$$dN = \frac{\hbar k}{mV}|f(\theta)|^2\,d\Omega \tag{9.61}$$

The differential cross section $\sigma(\theta)$ is obtained by dividing dN by $d\Omega$ and by the incident flux $\hbar k/mV$ (cf. (9.22)). Thus

$$\sigma(\theta) = |f(\theta)|^2$$

$$= \frac{1}{k^2}\left|\sum_{l=0}^{\infty}(2l+1)\,e^{i\delta_l}\sin\delta_l P_l(\cos\theta)\right|^2 \tag{9.62}$$

The total cross section σ is obtained by integrating (9.62) over all solid angles

$$\sigma = 2\pi\int_0^{\pi}\sigma(\theta)\sin\theta\,d\theta$$

$$= \frac{4\pi}{k^2}\sum_{l=0}^{\infty}(2l+1)\sin^2\delta_l \tag{9.63}$$

where we have used the expression

$$\int_0^{2\pi}\int_0^{\pi}P_l P_{l'}\sin\theta\,d\theta\,d\phi = \frac{2}{2l+1}\delta_{ll'} \tag{9.64}$$

($\delta_{ll'}$ being the Kronecker delta) which follows from (9.56) and the fact that the spherical harmonics are orthonormal.

Thus, if the phase shifts δ_l are known, the differential and total cross sections can be calculated using (9.62) and (9.63). The phase shifts are determined by the detailed size and shape of the scattering potential and their calculation is often a long and complicated process which is only practicable using numerical methods and computers. However, in some cases all but a few of the phase shifts are zero or negligibly small and expressions for the scattering cross sections are comparatively simple. This is particularly true in the case of the scattering of low energy particles when often the only significant contributions to (9.62) and (9.63) are from terms with $l = 0$. We can see why this is so from the following semi-classical argument. The partial waves correspond to particular values of the angular momentum L given by $\sqrt{l(l+1)}\hbar$; but classically a particle with this angular momentum and with linear momentum $\hbar k$ must pass the origin at a distance x such that $L = \hbar k x$ and therefore

$$kx = \sqrt{l(l+1)}$$

We can conclude, therefore, that if the range of the scattering potential is of order a (that is, $U(r)$ is zero or negligibly small if $r > a$) the scattering of the lth partial wave will not contribute unless

$$ka \geqslant \sqrt{l(l+1)} \qquad (9.65)$$

For sufficiently small k (i.e., for sufficiently low energies) (9.65) is satisfied only if $l = 0$ and this is then the only partial wave to be significantly scattered. This case is often referred to as s-wave scattering (cf. the terminology for $l = 0$ bound states set out in Chapter 3). It corresponds to a spherically symmetric wave and therefore isotropic scattering, in agreement with the result obtained earlier using the Born approximation in the special case where $ka \ll 1$. A rigorous quantum mechanical proof of the above involves investigating the form of the partial waves contributing to the incident plane wave near the origin, rather than in the asymptotic limit of large r. We shall not do this here, but simply quote the result that these are proportional to $(kr)^l$ in the limit $r \to 0$; in the case of small k the incident partial waves are therefore small over the volume of the scattering potential and no appreciable scattering results unless the condition (9.65) is satisfied.

We shall now discuss two examples of the application of the method of partial waves and in each case we shall confine our consideration to s-wave scattering.

Example 9.3 Scattering by hard spheres In this example we consider a scatterer whose radius is a and which cannot be penetrated by the incident particles. Thus the potential energy is given by

$$\left. \begin{array}{ll} U(r) = \infty & t \leqslant a \\ U(r) = 0 & r > a \end{array} \right\} \qquad (9.66)$$

It follows that the wave function must be zero for $r \leqslant a$ and we remember that it is continuous at the boundary $r = a$.

If we consider s-wave scattering only, it follows from (9.46) that the asymptotic form of the wave function (9.52) is actually a solution to the Schrödinger equation for all values of r greater than a so that the continuity equation becomes

$$e^{ika} - e^{i(ka + 2\delta_0)} = 0$$

and the phase shift δ_0 is therefore given by

$$\delta_0 = -ka \qquad (9.67)$$

The total cross section is obtained by substituting (9.67) into (9.63) to give

$$\sigma = \frac{4\pi \sin^2 ka}{k^2}$$

As the s-wave approximation is valid only in the limit $ka \ll 1$, we can write this as

$$\sigma \simeq 4\pi a^2 \tag{9.68}$$

Thus the scattering cross section is larger than the geometrical cross section by a factor of four.

Classically we would expect the scattering cross section for such a hard-sphere potential to be πa^2, so we have yet another example of how quantum mechanics produces results very different from those we would intuitively expect. In the next example, we shall see how a finite scattering potential can produce results that are even more dramatically different from classical expectations.

Example 9.4 Scattering by a potential well or step We now return to the problem previously treated using the Born approximation where the potential is given by

$$\left.\begin{array}{ll} U = 0 & r > a \\ U = U_0 & r \leqslant a \end{array}\right\} \tag{9.69}$$

As before, U_0 may be positive or negative, but now need not be small. We shall, however, consider explicitly only the case where the incident energy E is greater than U_0 and also confine our attention to s-wave scattering where $l = 0$ and $ka \ll 1$; it follows that our results will not be applicable to the repulsive case unless $U_0 \ll \hbar^2/2ma^2$.

In the region $r \leqslant a$ the general solution to the Schrödinger equation when $l = 0$ is readily seen to be

$$\frac{1}{r}(A' \sin k'r + B' \cos k'r) \tag{9.70}$$

where

$$k' = [2m(E - U_0)/\hbar^2]^{1/2} \tag{9.71}$$

and A' and B' are constants. However, the wave function must be finite everywhere, including the point $r = 0$, so the cosine term cannot exist and B' must be equal to zero. The $l = 0$ eigenfunction in the region $r > a$ again has the general form (9.52) and, if we apply the condition that both the wave function and its first spatial derivative must be continuous at $r = a$, we get

$$A[e^{-ika} - e^{i(ka + 2\delta_0)}] = A' \sin k'a$$

$$ikA[-e^{-ika} - e^{i(ka + 2\delta_0)}] = k'A' \cos k'a$$

If we divide the second of these equations by the first we get

$$\frac{ik[-e^{-ika} - e^{i(ka + 2\delta_0)}]}{e^{-ika} - e^{i(ka + 2\delta_0)}} = \frac{k' \cos k'a}{\sin k'a}$$

We can now multiply the numerator and denominator on the left-hand side by $\exp(i\delta_0)$ and express the resulting expressions in trigonometric form as

$$k \cot(ka + \delta_0) = k' \cot k'a$$

which leads, after a little manipulation, to

$$\cot \delta_0 = \frac{k \tan ka \tan k'a + k'}{k \tan k'a - k' \tan ka} \tag{9.72}$$

$$\simeq \frac{k' \cot k'a}{k(1 - k'a \cot k'a)} \tag{9.73}$$

where (9.73) is the limiting form of (9.72) when $ka \ll 1$. The scattering cross section for s-wave scattering is given by the $l = 0$ term in (9.63) as

$$\sigma = \frac{4\pi}{k^2} \sin^2 \delta_0$$

$$= \frac{4\pi}{k^2(1 + \cot^2 \delta_0)} \tag{9.74}$$

using standard trigonometric identities. Thus, given k, a, and U_0 we can calculate $\cot \delta_0$ and hence σ.

We note from (9.73) that if the magnitude of $k'a$ happens to be such that $k'a \cot k'a = 1$ (i.e., if $k'a = 4.49, 7.73, 10.9, \ldots$) $\cot \delta_0$ will be infinite, the cross section will be zero, and the incident particles will 'diffract past' the potential without being significantly scattered. Clearly this condition can be satisfied simultaneously with ka being small only if U_0 is negative—that is, if we are considering a potential well rather than a step. In such a case, the absence of scattering means that the wave function inside the well fits smoothly onto the plane wave outside. Clearly the potential well need not be square for this to occur, and the phenomenon has been observed experimentally in the case of the scattering of electrons by rare-gas atoms when it is known as the Ramsauer–Townsend effect: electrons of energy about 0.7 eV pass through helium gas without being significantly scattered.

The opposite extreme to the above case occurs when the parameters are such that $k'a$ is approximately equal to an odd multiple of $\pi/2$ so that $\cot k'a$ is small. We can then ignore the second term in the denominator of (9.73) and write the cross section as

$$\sigma = \frac{4\pi}{k^2 + k'^2 \cot^2 k'a} \tag{9.75}$$

and we note that, as ka and $\cot k'a$ are both much less than one, the cross section (9.75) is much greater than the geometrical cross section πa^2. This case is known as *resonant scattering*. It is interesting to compare the condition for such an s-wave resonance with the equation governing the energy levels of a particle in a spherical potential well (cf. Prob. 3.5) which is

$$-k'' \cot k''a = (-2mE_b/\hbar^2)^{1/2} \tag{9.76}$$

where

$$k'' = [2m(U_0 - E_b)/\hbar^2]^{1/2}$$

and E_b is the binding energy of the system. In the case where E_b is small, k'' is nearly equal to k' and (9.76) is approximately equivalent to the resonance condition. We can conclude, therefore, that s-wave resonant scattering with large cross section is to be expected if there is a bound state of the system particle-plus-scatterer whose energy level lies just below the top of the well; the cross section (9.75) can then be expressed as

$$\sigma \simeq \frac{2\pi\hbar^2}{m(E + E_b)} \tag{9.77}$$

We note that as the energy E of the incident particles is decreased, the cross section approaches a limiting value determined by the binding energy. Clearly the above conditions will also hold and resonance scattering will also occur if U_0 is just a little smaller than would be required for there to be zero-energy bound state. In this case the quantity E_b measured from such scattering experiments is interpreted as the energy associated with a 'virtual' energy level. As with the Ramsauer–Townsend effect, we expect these results to be quite general and independent of the detailed shape of the well; the large phase shift can be explained semi-classically if we imagine the incident particle to be 'trapped' in the well for a time before it is re-emitted, and the closer the energy of the particle is to that of the bound state, the greater will be the probability of trapping. An example of resonant scattering is that of low-energy neutrons from hydrogen where the s-wave cross section of 20.4×10^{-28} m^2 is more than one hundred times greater than would be expected classically, given the range of the neutron–proton interaction (2×10^{-15}); and indeed the deuteron (which consists of a proton and neutron bound together) does have an energy level with a very small binding energy.

PROBLEMS

9.1 Particles of energy E move in one dimension towards a potential barrier of height U_0 and width a. Show that if E is greater than U_0 the probability of scattering P is given by

$$P = \left[1 + \frac{4E(E - U_0)}{U_0^2 \sin^2 k_2 a}\right]^{-1}$$

where $k_2 = [2m(E - U_0)/\hbar^2]^{1/2}$, and obtain a corresponding expression for the transmission probability.

9.2 Show that the scattering probability derived in Prob. 9.1 is zero if $k_2 a = n\pi$ where n is an integer and has a maximum value if the incident energy is such that $k_2 a \cot k_2 a = 2 - U_0/E$.

Compare this situation and that set out in Prob. 9.1 with those applying in the Ramsauer–Townsend effect and in s-wave resonant scattering.

9.3 Obtain expressions for the scattering and transmission probabilities for the system described in Prob. 9.1, but now assume E to be less than. U_0. Use your answers to all three questions to draw graphs of the scattering probability as a function of E for values of E between 0 and $5U_0$ in the cases where $U_0 a^2 = \pm 10\hbar^2/m$.

9.4 Show that in one dimension the scattering probability calculated using the Born approximation is P' where

$$P' = \frac{m^2}{\hbar^4 k^2} \left| \int_{-\infty}^{\infty} U(x) e^{-2ikx} dx \right|^2$$

Use this expression to calculate the scattering probability for the system described in Prob. 9.1 and show that it is equivalent to the expression given there in the limit $E \gg U_0$.

9.5 Particles of momentum $\hbar k_0$ travel along the z axis towards a three-dimensional rectangular potential well of depth U_0 and dimensions $2a \times 2b \times 2c$ where a, b, and c are parallel to the x, y, and z axes respectively. Use the Born approximation to show that, if the incident energy is much greater than U_0, the differential scattering cross section in the direction (k_x, k_y, k_z) is equal to

$$\frac{16m^2 U_0^2 \sin^2(k_x a) \sin^2(k_y b) \sin^2[(k_z - k_0)c]}{\pi^2 \hbar^4 k_x^2 k_y^2 (k_z - k_0)^2}$$

9.6 Use the Born approximation to estimate the differential scattering cross section when the scattering potential is spherically symmetric and has the form Ar^{-2} where A is a constant.

Hint: $\int_0^x x^{-1} \sin x \, dx = \pi/2$.

9.7 Show that the expression for s-wave scattering by a spherical potential well—see Eqs (9.72) and (9.74)—goes over to that obtained from the Born approximation (cf. 9.39) in the limit where $E \gg U_0$ and $ka \ll 1$.

Hint: do not use (9.73), but obtain a limit of (9.72) when k and k' are both small.

9.8 Show that the phase shift of the $l = 0$ partial wave in the case of scattering from a potential step, whose height U_0 is greater than the energy E of the incident particles, is given by Eq. (9.72) with $\tan k'a$ replaced by $\tanh k'a$ wherever it appears. Show that in the limit $U_0 \to \infty$ this result is consistent with that obtained earlier in the case of scattering by hard spheres.

9.9 Show that the s-wave cross section for very low-energy neutrons scattered by protons is about 2.4×10^{-28} m^2, on the assumption that the binding energy of the deuteron is -2.23 MeV (remember to use reduced mass). This is considerably smaller than the observed cross section of 20.4×10^{-28} m^2 because of the spin dependence of the neutron–proton interaction. In the ground state of the deuteron the z components of the two spins are parallel, but this is true for only three-quarters of all scattering events. Show that the experimental cross section can be reproduced if the state with zero total spin has a virtual level with energy about 70 keV.

MANY-PARTICLE SYSTEMS

Although many of the principles of quantum mechanics can be adequately illustrated by considering systems that consist of only one particle subject to external forces, there are a number of important phenomena that are manifest only in systems containing two or more particles, and these will be the subject of the present chapter. We shall begin with some general statements and then consider the case of two interacting particles subject to no external forces, when we shall find that the problem can be separated into one describing the behaviour of the centre of mass of the system and another describing the relative motion. We shall then consider the case of two non-interacting particles and show that the particles can usually be treated independently as would be expected. In the case of indistinguishable particles, however, we shall find that a symmetry is imposed on the wave function which ensures that the behaviour of such particles is coupled even when they do not interact. These r ..its will then be illustrated by discussing the energy levels and optical spectra of the helium atom, starting with the approximation that the two electrons do not interact and extending the discussion to the realistic case using perturbation theory. We shall then briefly introduce the problem of interacting ystems containing more than two identical particles, and close the chapter with a consideration of the effects of particle indistinguishability on scattering theory.

10.1 GENERAL CONSIDERATIONS

If we have a system containing N particles, it will be described by a wave function that is in general a function of the positions and spins of all the particles and of time. Ignoring spin for the present, we write this wave function as $\Psi(\mathbf{r}_1, \mathbf{r}_2, \ldots, \mathbf{r}_N, t)$ and we can straightforwardly extend the probabilistic interpretation set up in the earlier chapters for a single particle, so that

$$|\Psi(\mathbf{r}_1, \mathbf{r}_2, \ldots, \mathbf{r}_N, t)|^2 \, d\tau_1 \, d\tau_2, \ldots, d\tau_N \tag{10.1}$$

is the probability that particle 1 will be found in the element of volume $d\tau_1$ in the vicinity of $\mathbf{r}_1$ simultaneously with particle 2 being found within $d\tau_2$ in the vicinity of $\mathbf{r}_2$, etc. Probability distributions for other dynamical quantities can be derived from Ψ by the general procedure described in Chapter 4, given the form of the appropriate operators and assuming that the corresponding eigenvalue equations can be solved. In general, integrals occurring in the theoretical expressions are with respect to all the coordinates of all the particles in the system.

The operators representing measurable quantities may be specific to particular particles or may represent global properties of the system. Thus the operator representing the x component of the momentum of the ith particle is $\hat{P}_{xi} = -i\hbar \partial/\partial x_i$, while that representing the x component of the total momentum of a system of N particles is

$$\hat{P}_x = \sum_{i=1,N} \hat{P}_{xi} = -i\hbar \sum_{i=1,N} \partial/\partial x_i \tag{10.2}$$

As usual, we shall be particularly interested in the energy eigenvalues and eigenfunctions, and the Hamiltonian operator representing the total energy of a system of N particles has the following general form (omitting possible spin-dependent terms)

$$\hat{H} = -\sum_{i=1,N} \frac{\hbar^2}{2m_i} \nabla_i^2 + V(\mathbf{r}_1, \mathbf{r}_2, \ldots, \mathbf{r}_N) \tag{10.3}$$

where m_i is the mass of the ith particle, ∇_i is the vector operator differentiating with respect to the coordinates of the ith particle, and V is the potential energy of the system which in general will include contributions from external forces along with the energy associated with the interactions between the particles. The eigenvalue equation corresponding to the Hamiltonian (10.3) is therefore a partial differential equation containing $3N$ variables. We saw in the early chapters how difficult the solution of a system containing only the variables associated with a single particle could be, so it is not surprising that exact solutions to the many-body problem are possible only in particular simple cases. In most of the ensuing discussion we shall refer to systems containing only two particles, because these turn out to illustrate most of the principal features of many-body systems, and we shall return briefly to problems concerning more than two particles in Sec. 10.6.

10.2 ISOLATED SYSTEMS

In this section we consider a system of two particles that are not subject to any external forces so that the potential energy depends only on the relative position of the particles and can be written as $V(\mathbf{r}_1 - \mathbf{r}_2)$. The Hamiltonian is then

$$\hat{H} = -\frac{\hbar^2}{2m_1} \nabla_1^2 - \frac{\hbar^2}{2m_2} \nabla_2^2 + V(\mathbf{r}_1 - \mathbf{r}_2) \qquad (10.4)$$

We now change variables from the particle positions $\mathbf{r}_1$ and $\mathbf{r}_2$ to those of the centre of mass of the system $(\mathbf{R})$ and the relative position $\mathbf{r}$. That is

$$\left.\begin{aligned} \mathbf{R} &= \frac{m_1\mathbf{r}_1 + m_2\mathbf{r}_2}{m_1 + m_2} \\ \mathbf{r} &= \mathbf{r}_1 - \mathbf{r}_2 \end{aligned}\right\} \qquad (10.5)$$

The vector differential operators ∇_1 and ∇_2 can be expressed in terms of the quantities ∇_R and ∇_r which correspond to the variables $\mathbf{R}$ and $\mathbf{r}$ respectively by

$$\left.\begin{aligned} \nabla_1 &= \frac{m_1}{m_1 + m_2} \nabla_R + \nabla_r \\ \nabla_2 &= \frac{m_2}{m_1 + m_2} \nabla_R - \nabla_r \end{aligned}\right\} \qquad (10.6)$$

Substituting from (10.5) and (10.6) into (10.4) we get

$$\hat{H} = -\frac{\hbar^2}{2M} \nabla_R^2 - \frac{\hbar^2}{2\mu} \nabla_r^2 + V(\mathbf{r}) \qquad (10.7)$$

where M is the total mass $(m_1 + m_2)$ and μ is the 'reduced mass', $m_1 m_2/(m_1 + m_2)$. Clearly the energy eigenvalue equation can now be separated and the eigenfunction written as $U(\mathbf{R})u(\mathbf{r})$ where

$$-\frac{\hbar^2}{2M} \nabla_R^2 U = E_R U \qquad (10.8)$$

and

$$\left[-\frac{\hbar^2}{2\mu} \nabla_r^2 + V \right] u = E_r u \qquad (10.9)$$

the total energy being equal to $E_R + E_r$. Thus the problem has been separated into one concerning only the centre of mass of the system and another which describes the behaviour of a particle of mass μ under the influence of a potential $V(\mathbf{r})$. This justifies the procedure used when we obtained the energy levels of the hydrogen atom in Chapter 3 (see footnote to p. 53) and when we considered the scattering of neutrons by protons in Chapter 9 (Problem 9.9).

The separation described above is also possible if the system is not isolated, so long as the external forces can be considered as acting on the centre of mass of the system, because the total potential then has the form $V_R(\mathbf{R}) + V_r(\mathbf{r})$ and Eq. (10.3) can be separated into two equations, one in $\mathbf{R}$ and the other in $\mathbf{r}$. The motion of the centre of mass can also be separated out in the case of a similar system containing more than two particles, although the residual equations describing the internal motion do not now have such a simple form. This explains why the behaviour of the centre of mass of a composite system can be treated without having to consider the detailed behaviour of its component particles. For example, provided the thermal energy is not sufficient to cause electronic excitations, the behaviour of an atom in a gas can be described without taking its internal structure into account. Similarly, the internal structure of the nucleus can be ignored when discussing most of the properties of an atom, while a 'fundamental' particle such as a proton can be considered as a point particle unless very high energy interactions, disturbing its internal structure, are involved.

10.3 NON-INTERACTING PARTICLES

We now turn to a case which is more or less the opposite of that just discussed, and consider two particles which may be subject to external forces but which do not interact with each other. The potential can then be written as $V_1(\mathbf{r}_1) + V_2(\mathbf{r}_2)$ and the Hamiltonian becomes

$$-\frac{\hbar^2}{2m_1} \nabla_1^2 - \frac{\hbar^2}{2m_2} \nabla_2^2 + V_1(\mathbf{r}_1) + V_2(\mathbf{r}_2) \qquad (10.10)$$

If we now write an energy eigenfunction in the form $u_1(\mathbf{r}_1)u_2(\mathbf{r}_2)$ we can separate the variables $\mathbf{r}_1$ and $\mathbf{r}_2$ and get the following eigenvalue equations

$$-\frac{\hbar^2}{2m_1} \nabla_1^2 u_1 + V_1 u_1 = E_1 u_1 \qquad (10.11)$$

$$-\frac{\hbar^2}{2m_2} \nabla_2^2 u_2 + V_2 u_2 = E_2 u_2 \qquad (10.12)$$

where E_1 and E_2 are the energies associated with the separate particles and the total energy is

$$E = E_1 + E_2 \qquad (10.13)$$

Thus we can apparently treat the particles as completely independent of each other, as would be expected if indeed they do not interact. However, this apparently obvious conclusion is not valid in the particular case where the particles are identical, and we shall discuss this situation in more detail in the next section.

10.4 INDISTINGUISHABLE PARTICLES

Identical particles are often referred to as *indistinguishable* in order to emphasize the fact that they cannot be distinguished by any physical measurement. This implies that an operator representing any physical measurement on the system must remain unchanged if the labels assigned to the individual particles are interchanged. Thus, if we write the Hamiltonian of two indistinguishable (although now not necessarily non-interacting) particles as $\hat{H}(1, 2)$, we must have

$$\hat{H}(1, 2) = \hat{H}(2, 1) \tag{10.14}$$

We now define a 'particle interchange operator' $\hat{P}_{12}$ such that if this operates on any function or operator that depends on the variables describing two particles it has the effect of interchanging the labels.

Thus we have

$$\hat{P}_{12}\hat{H}(1, 2) = \hat{H}(2, 1) = \hat{H}(1, 2) \tag{10.15}$$

using (10.14). Now let $\psi(1, 2)$ be a wave function (not necessarily an energy eigenfunction) describing the two particles. The wave function is not a physical quantity and so need not be invariant when operated on by $\hat{P}_{12}$. However we can write

$$\hat{P}_{12}\hat{H}(1, 2)\psi(1, 2) = \hat{H}(2, 1)\psi(2, 1)$$

$$= \hat{H}(1, 2)\hat{P}_{12}\psi(1, 2)$$

Hence

$$(\hat{P}_{12}\hat{H}(1, 2) - \hat{H}(1, 2)\hat{P}_{12})\psi(1, 2) = 0 \tag{10.16}$$

that is

$$[\hat{P}_{12}, \hat{H}(1, 2)] = 0 \tag{10.17}$$

because (10.16) is true whatever the form of ψ. Thus the particle interchange operator and the Hamiltonian commute which means that they are compatible and have a common set of eigenfunctions. Now, because the particles are indistinguishable, not only the Hamiltonian but any operator representing a physical property of the system must be symmetric with respect to particle interchange and must therefore commute with $\hat{P}_{12}$. Thus whatever measurement is made on the system, the resulting wave function will be an eigenfunction of $\hat{P}_{12}$, and no loss of generality is involved if we assume that the wave function always has this property.

If $\psi(1, 2)$ is to be an eigenfunction of $\hat{P}_{12}$, it must be a solution of the eigenvalue equation

$$\hat{P}_{12}\psi(1, 2) = p\psi(1, 2) \qquad (10.18)$$

where p is the corresponding eigenvalue. But the left-hand side of (10.18) is by definition equal to $\psi(2, 1)$. That is,

$$\psi(2, 1) = p\psi(1, 2) \qquad (10.19)$$

We can now operate on both sides of (10.19) with $\hat{P}_{12}$ to get

$$\psi(1, 2) = \hat{P}_{12}\psi(2, 1) = p\hat{P}_{12}\psi(1, 2)$$
$$= p^2\psi(1, 2)$$

That is,

$$p = \pm 1 \qquad (10.20)$$

and

$$\psi(1, 2) = \pm\psi(2, 1) \qquad (10.21)$$

We therefore conclude that any physically acceptable wave function representing two identical particles must be either symmetric or antisymmetric with respect to an interchange of the particles. This property is readily extended to a system containing more than two particles when the many-body wave function must be either symmetric or antisymmetric with respect to the interchange of any pair of particles. Moreover, we have seen that the operators representing physical measurements all commute with $\hat{P}_{12}$ so that, once a system is in one of these eigenstates, it can never make a transition to the other.

It follows from the above that every particle belongs to one of two classes depending on whether the wave function representing a number of them is symmetric or antisymmetric with respect to particle exhange. Particles with symmetric wave functions are known as *bosons* while those whose wave functions are antisymmetric are called *fermions*. This symmetry with respect to particle interchange turns out to be closely connected with the value of the total spin of the particle: bosons always have an integral total-spin quantum number (e.g., the α particle and pion which have spin-zero and the deuteron which is a spin-one particle) while fermions always have half-integral spin (e.g., the electron, proton, neutron, and neutrino which are all spin-half). This simple one-to-one correspondence between the total-spin quantum number and the interchange symmetry can be shown to be a necessary consequence of relativistic quantum field theory, but such an argument is beyond the scope of this book.

Non-Interacting Indistinguishable Particles

The results obtained so far in this section apply to any system of indistinguishable particles, whether or not there is any interaction between them. We can now combine these with the particular properties of non-interacting systems discussed earlier to find the form of the energy eigenfunction of two indistinguishable non-interacting particles. Referring to Eqs (10.10) we see that, if the particles are indistinguishable, m_1 must equal m_2 and $V_1(\mathbf{r})$ must be the same as $V_2(\mathbf{r})$. Thus Eqs (10.11) and (10.12) are now identical and have the same set of eigenvalues and eigenfunctions. The total energies of the states with eigenfunctions $u_1(1)u_2(2)$ and $u_1(2)u_2(1)$ are therefore the same, and any linear combination of these is also an eigenfunction with the same eigenvalue. We must therefore form linear combinations of these products which have the appropriate symmetry with respect to particle exchange.

Considering bosons first, the wave function must be symmetric with respect to particle exchange so we must have

$$\psi(1,2) = 2^{-1/2}[u_1(1)u_2(2) + u_1(2)u_2(1)] \qquad (10.22)$$

where the factor $2^{-1/2}$ ensures that the wave function is normalized. In the special case where both particles are associated with the same single-particle state, that is where u_1 is the same as u_2, (10.22) becomes (with a slight change of normalization factor)

$$\psi(1,2) = u_1(1)u_1(2) \qquad (10.23)$$

Turning now to fermions, we must form an antisymmetric linear combination of the degenerate functions to get

$$\psi(1,2) = 2^{-1/2}[u_1(1)u_2(2) - u_1(2)u_2(1)] \qquad (10.24)$$

and this time the wave function in the special case where u_1 is the same as u_2 becomes

$$\psi(1,2) = 2^{-1/2}[u_1(1)u_1(2) - u_1(1)u_1(2)]$$
$$= 0$$

so that the wave function now vanishes identically implying that such a state of the system does not exist. This result is known as the *Pauli exclusion principle* and is sometimes expressed by stating that no two fermions can be the same state. This can be a very useful form of the exclusion principle provided we realize that if the energy eigenfunction is to be antisymmetric it must have the form (10.24) and therefore a particular particle cannot actually be identified with a particular single-particle function u_1 or u_2. The exclusion principle actually states that each single-particle eigenfunction can be used only once in constructing products, linear combinations of which form the total wave functions. The reason the simpler form can be used is that we are often interested only in the total energy of a system of non-interacting particles. Because of degeneracy, we will get the correct value for the energy if we use simple product wave functions of the form $u_1(1)u_2(2)$ provided we apply the exclusion principle to exclude any products in which both particles are assigned to the same state. We used this procedure in our discussion of the properties of free electrons in solids in Chapter 7 where we found that the differences between insulators and metals arise from the fact that all states up to an energy gap are full in the former case, but not the latter, and it turns out that these simple results remain valid even when inter-electronic interactions are included. Similar principles can be applied to electrons in atoms to give a qualitative account of many of the chemical properties of the elements and the periodic table. For example, in the ground state of the lithium atom, two of the three electrons occupy the lowest ($1s$) energy level with opposite spin and the third is in the higher-energy $2s$ state, while in the sodium atom, ten electrons fill all the levels with $n = 1$ and $n = 2$, leaving one in the $3s$ level. When all the states with the same n value are full, the electrons in them have comparatively little effect on the physical and chemical properties of the element; it follows that these are largely determined by the one outer electron only and are therefore very similar in the two cases.

If, however, we are interested in properties other than the total energy of a system of non-interacting particles, or if we wish to take interparticle interactions into account, we have to use a wave function which is antisymmetric with respect to particle interchange and we must abandon any idea of a particle being associated with a particular state. We shall consider an example of such a system containing two interacting particles in the next section when we discuss the helium atom, and we shall return to the conceptual problems associated with the non-separability of identical particles in Chapter 11.

10.5 THE HELIUM ATOM

The helium atom consists of two electrons and a doubly positively charged nucleus. Throughout our discussion we shall make the approximation that the mass of the nucleus is infinitely greater than that of an electron so that the problem can be treated as that of two electrons moving in a potential. For the moment we shall also ignore spin-orbit coupling so that the Hamiltonian of the system is

$$\hat{H} = -\frac{\hbar^2}{2m}\nabla_1^2 - \frac{\hbar^2}{2m}\nabla_2^2 - \frac{2e^2}{4\pi\varepsilon_0 r_1} - \frac{2e^2}{4\pi\varepsilon_0 r_2} + \frac{e^2}{4\pi\varepsilon_0 r_{12}} \qquad (10.25)$$

where $r_{12} \equiv |\mathbf{r}_1 - \mathbf{r}_2|$ is the separation of the two electrons. We note that, as expected, $\hat{H}$ is symmetric with respect to the interchange of the labels on the identical electrons. In the discussion that follows, we shall first ignore the last term in (10.25) which represents the interelectronic interaction so that we can consider the problem as a non-interacting one. The effects of the interaction term will then be treated by perturbation theory and finally spin-orbit interactions will be considered.

We saw earlier that in the non-interacting limit, the energy eigenfunctions can be expressed as products of single-particle eigenfunctions which in general will depend on both the particle position and its spin. In the case of helium it follows that, if the interaction is ignored, the spatial parts of the single-particle eigenfunctions are solutions to the equation

$$\left(-\frac{\hbar^2}{2m} \nabla_1^2 - \frac{2e^2}{4\pi\varepsilon_0 r_1} \right) u_n = E u_n \qquad (10.26)$$

and are therefore the same as the hydrogenic eigenfunctions described in Chapter 3 and referred to several times since. It will also be important to include the spin part of the wave function and we represent this by α or β depending on whether the z component of spin is positive or negative respectively. If we now consider states where the spatial parts of the single-particle eigenfunctions are u_1 and u_2, and if we include all possible values of the spin, we obtain the following eight products, all of which are energy eigenfunctions with the same eigenvalue in the non-interacting limit:

$$\left.\begin{array}{ll}
u_1(\mathbf{r}_1)\alpha(1)u_2(\mathbf{r}_2)\alpha(2) & u_2(\mathbf{r}_1)\alpha(1)u_1(\mathbf{r}_2)\alpha(2) \\[4pt]
u_1(\mathbf{r}_1)\alpha(1)u_2(\mathbf{r}_2)\beta(2) & u_2(\mathbf{r}_1)\beta(1)u_1(\mathbf{r}_2)\alpha(2) \\[4pt]
u_1(\mathbf{r}_1)\beta(1)u_2(\mathbf{r}_2)\alpha(2) & u_2(\mathbf{r}_1)\alpha(1)u_1(\mathbf{r}_2)\beta(2) \\[4pt]
u_1(\mathbf{r}_1)\beta(1)u_2(\mathbf{r}_2)\beta(2) & u_2(\mathbf{r}_1)\beta(1)u_1(\mathbf{r}_2)\beta(2)
\end{array}\right\} \qquad (10.27)$$

We shall now construct antisymmetric functions by taking appropriate linear combinations of the products listed in (10.27). This can be done in a number of ways, but it will turn out to be an advantage if they are each expressed as a product of a spatially dependent and a spin-dependent part; in this case, if the total wave function is to be antisymmetric, either the spin-dependent part must be antisymmetric and the spatial part symmetric or vice versa. The four functions which can be constructed consistently with these requirements are then:

$$\left.\begin{array}{l}
\dfrac{1}{\sqrt{2}}[u_1(\mathbf{r}_1)u_2(\mathbf{r}_2) - u_1(\mathbf{r}_2)u_2(\mathbf{r}_1)]\alpha(1)\alpha(2) \\[16pt]
\dfrac{1}{\sqrt{2}}[u_1(\mathbf{r}_1)u_2(\mathbf{r}_2) - u_1(\mathbf{r}_2)u_2(\mathbf{r}_1)]\beta(1)\beta(2) \\[16pt]
\dfrac{1}{\sqrt{2}}[u_1(\mathbf{r}_1)u_2(\mathbf{r}_2) - u_1(\mathbf{r}_2)u_2(\mathbf{r}_1)]\dfrac{1}{\sqrt{2}}[\alpha(1)\beta(2) + \alpha(2)\beta(1)] \\[16pt]
\dfrac{1}{\sqrt{2}}[u_1(\mathbf{r}_1)u_2(\mathbf{r}_2) + u_1(\mathbf{r}_2)u_2(\mathbf{r}_1)]\dfrac{1}{\sqrt{2}}[\alpha(1)\beta(2) - \alpha(2)\beta(1)]
\end{array}\right\} \qquad (10.28)$$

It should be noted that the spin parts of the functions are eigenfunctions of the operators representing the total spin of the two particles and of their total z component (cf. Sec. 6.4): the first three functions listed have their total-spin quantum number S equal to one and the quantum number associated with the z component m_s has the values plus one, minus one, and zero respectively; the fourth function has zero spin, that is $S = m_s = 0$. Thus in the first three states the electron spins are aligned as nearly parallel as is allowed by the quantization rules for angular momentum, while in the fourth they are anti-parallel. We also note that in the special case where the spatial parts of the single-particle functions are identical, that is, where $u_1 \equiv u_2$, the only state allowed by the exclusion principle is the one with $S = 0$, and the others are identically zero as expected.

We shall now consider how the energies of the above states are affected by the inclusion of the interelectronic electrostatic interaction and by the spin-orbit coupling.

Interelectronic Interactions

We consider first the effects of the electrostatic interactions between the electrons. Ignoring the spin parts of (10.28) for the moment, we apply perturbation theory to two degenerate states whose unperturbed eigenfunctions are

$$\left. \begin{array}{c} v_{01} = \dfrac{1}{\sqrt{2}} \left[u_1(\mathbf{r}_1)u_2(\mathbf{r}_2) - u_1(\mathbf{r}_2)u_2(\mathbf{r}_1) \right] \\[4mm] \text{and} \qquad v_{02} = \dfrac{1}{\sqrt{2}} \left[u_1(\mathbf{r}_1)u_2(\mathbf{r}_2) + u_1(\mathbf{r}_2)u_2(\mathbf{r}_1) \right] \end{array} \right\} \qquad (10.29)$$

where the perturbation is

$$H' = \frac{e^2}{4\pi\varepsilon_0 r_{12}} \qquad (10.30)$$

Following the standard procedure described in Chapter 7 we form the matrix elements of $\hat{H}'$ and find that the off-diagonal element H'_{12} is zero, so that there is no mixing of the two states and v_{01} and v_{02} are therefore the correct zero-order eigenfunctions to be used when calculating the effects of the perturbation. The first-order changes (E_{11} and E_{12}) to the energies of the two states are then equal to the diagonal elements of the perturbation matrix and we have

$$\left. \begin{array}{l} E_{11} = H'_{11} = C - X \\ E_{12} = H'_{22} = C + X \end{array} \right\} \qquad (10.31)$$

where
$$C = \int \int u_1^*(\mathbf{r}_1) u_2^*(\mathbf{r}_2) \frac{e^2}{4\pi\varepsilon_0 r_{12}} u_1(\mathbf{r}_1) u_2(\mathbf{r}_2) \, d\tau_1 \, d\tau_2$$

and
$$X = \int \int u_1^*(\mathbf{r}_1) u_2^*(\mathbf{r}_2) \frac{e^2}{4\pi\varepsilon_0 r_{12}} u_1(\mathbf{r}_2) u_2(\mathbf{r}_1) \, d\tau_1 \, d\tau_2$$

$$(10.32)$$

The integral C is sometimes referred to as the 'Coulomb energy' because it is equivalent to that of the classical electrostatic interaction between two continuous charge distributions whose densities are $|u_1|^2$ and $|u_2|^2$ respectively, while X is often referred to as the 'exchange energy' because the integrand contains terms where the numbers labelling the electrons have been 'exchanged'. Very little physical significance should be attached to these designations, however, as the integrals arise simply as a result of applying perturbation theory to the system, and the only physically significant quantity being calculated is the total energy whose value is equal to either the sum or difference of the two terms.

The quantities C and X can be evaluated given the form of the one-electron eigenfunctions u_1 and u_2. The procedure is reasonably straight forward in principle, but rather complex and tedious in detail. Accordingly we shall not describe such calculations here, but concentrate on the qualitative significance of the expressions for the energy eigenvalues, although we shall briefly discuss the application of Eqs (10.32) to the calculation of the ground-state energy towards the end of this section.

We first note that, whereas there are four degenerate eigenfunctions (10.28) in the absence of a perturbation, the inclusion of the electrostatic interaction has split this system into a 'triplet' of three degenerate states all of which have total-spin quantum number equal to one, and a spin-zero 'singlet' state. Thus the energies of the states where the spins are parallel are different from those where the spins are anti-parallel and the system therefore behaves as though the spins were strongly coupled. It is important to remember that this coupling results from the requirement that the wave function be antisymmetric and is quite independent of the interaction between the magnetic dipoles associated with the spins which is very much weaker. It turns out that the exchange integral X is generally positive so that the triplet states have lower energy than the corresponding singlets. Nevertheless, the ground state of helium is a singlet, because in this case both electrons occupy the same ($1s$) orbital and no corresponding triplet state exists. Another type of physical system which exhibits a similar coupling of spins is a ferromagnet such as iron: in this case, however, the exchange interaction between electrons associated with neighbouring atoms leads to a negative X and hence is such as to produce a triplet ground state. The magnetic moments associated with the atomic spins therefore all point in the same direction, leading to a large magnetic moment overall.

Returning to the helium atom, the coupling of the spins via the exchange energy means that the spin-orbit interaction is between the total spin and the total orbital angular momentum, whose magnitude is determined by the quantum numbers of the particular single-particle functions u_1 and u_2. Given these quantum numbers, the splitting due to spin-orbit coupling can be calculated by the methods discussed in Chapter 6, as can the response of the system to weak and strong magnetic fields. Helium is therefore an example of Russell–Saunders coupling which was discussed briefly at the end of Chapter 6 where we mentioned that it applied most usefully to atoms of low atomic number. In the case of heavy atoms, the magnetic interaction between the spin and orbital angular momenta of the individual electrons turns out to be stronger than the exchange interaction and we get *j-j* coupling.

The fact that the spatial part of a singlet wave function is symmetric, while that of a triplet is antisymmetric, means that electric dipole transitions between any members of these two sets of states are forbidden. This is because the electric dipole operator is a function of the electron positions only and, like all physical operators, is symmetric with respect to particle interchange so that the matrix elements connecting the singlet and triplet states are of the form

$$\iint \psi_s^*(1,2)\hat{Q}_s(1,2)\psi_a(1,2)\, d\tau_1\, d\tau_2 = -\iint \psi_s^*(2,1)\hat{Q}_s(2,1)\psi_a(2,1)\, d\tau_1\, d\tau_2$$

$$(10.33)$$

where the subscripts s and a signify symmetry and antisymmetry respectively and $\hat{Q}_s$ is any symmetric operator. But the labels on the variables of integration have no significance, so the integrals on each side of (10.33) must be identical, which can be true only if they both vanish. The same is true for any perturbation that is a function of either the electron positions or of the spins only. It follows that singlet to triplet transitions, and vice versa, occur only as a result of collisions between atoms, where the property conserved is the total antisymmetry of the wave function representing all the electrons in both atoms, rather than that of each atom separately. Because transitions between singlet and triplet states occur so rarely, they were not observed at all in the early days of spectroscopy, and at one time helium was thought to be a mixture of two gases: 'parahelium' with the singlet spectrum and 'orthohelium' with that of the triplet.

More recently the above properties were used in the construction of the helium–neon laser. When an electric discharge is passed through helium gas, it causes many of the atoms to be ionized and, when they subsequently recombine, there is an appreciable probability of some of the atoms being in one of the excited triplet states. These then decay, emitting photons, until they reach the lowest energy triplet state where they remain because a further transition to the singlet ground state is forbidden. It turns out that the excitation energy of the lowest triplet state of helium is close to that of one of the excited states of neon so that in a mixture of the two gases there is an appreciable probability of neon atoms being excited into this state by collisions with appropriate helium atoms. If the partial pressures of the gases in such a mixture are right, an inverted population can be generated in which more neon atoms are in this excited state than are in another state which has lower energy. Radiation whose frequency matches the energy difference between the two neon states can therefore be amplified by stimulated emission and laser action ensues, which can be maintained continuously by passing a suitable discharge through the mixture, thereby replenishing the number of triplet helium atoms.†

We shall complete our discussion of the helium atom by considering some of the more quantitative features of the energy levels. Figure 10.1 shows an energy-level diagram in which the experimentally measured energies of the states are referred to an origin that corresponds to the energy required to just remove one electron from the atom, leaving the other in the ground state of the remaining He^+ ion. The energy levels of the hydrogen atom derived in Chapter 3 are also included for comparison. We see that, on this scale, the ground-state energy of helium is very much lower than that of hydrogen and that this is also true to a lesser extent of the first and second excited states. On the other hand, states with high values of the principal quantum number n (i.e., greater than about five) have energies that are very little different from the corresponding hydrogen-atom values, being nearly independent of the orbital angular momentum quantum number l. This last point is consistent with the fact that helium, like hydrogen, exhibits a first-order Stark effect (cf. Sec. 7.2).

We can account for all the above features using the theory developed earlier. The first point is that all the bound states of the helium atom involve linear combinations of products in which one of the one-electron functions is the lowest energy ($1s$) orbital, because states where both electrons are assigned to excited orbitals turn out to have such a high energy that they cannot exist as bound states. Accordingly, the quantity referred to above as the principal quantum number n denotes the orbital other than the $1s$ which is involved in the description of the state. We first consider highly excited states and note that the one-electron orbitals with high values of n have appreciable magnitude only at large distances from the nucleus, while the $1s$ eigenfunction is significant only at small values of r. If we now examine the expressions for the Coulomb and exchange energies (10.32) we see that, for a highly excited state, the exchange energy is very small because u_1 and u_2 occupy almost entirely different regions of space. The singlet and triplet states can therefore be treated as degenerate, which means that the energy can be estimated by assigning one electron to each of the orbitals u_1 and u_2 in the same way as in the non-interacting case discussed earlier. The total energy of such a state would then be equal to the sum of the energies of the

---

†A more detailed account of the operation of the helium–neon laser can be found in, for example, L. Allen, *Essentials of Lasers*, Pergamon, Oxford, 1969.

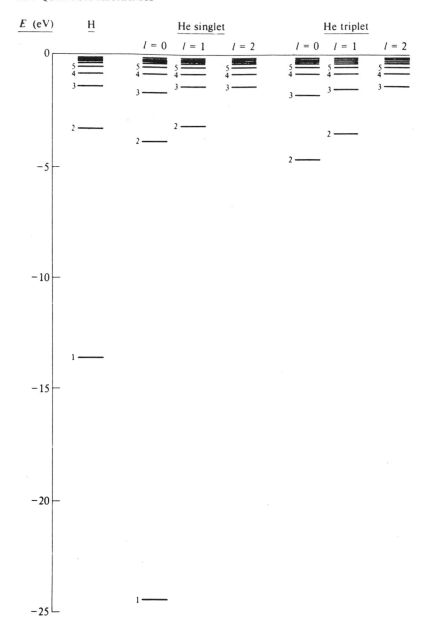

Figure 10.1 The energies of the singlet and triplet states of the helium atom compared with the energy levels of the hydrogen atom. The number beside each energy level is the principal quantum number n. Each triplet state is actually three closely spaced levels split by the spin-orbit interaction.

ground and appropriately excited states of the He^+ ion, along with the Coulomb energy which represents the mean electrostatic interaction between electrons assigned to these two orbitals. This result still has the limitations of first-order perturbation theory, and a better approximation is obtained by assuming that the particles can be independently assigned to each single-particle orbital but that the form of each orbital is affected by the presence of the other electron. Thus the outer electron moves in a potential due to the nucleus and the charge distribution of the spherically symmetric inner orbital, which is equivalent to that of a single positive charge $(+e)$ at the nucleus. The outer electron therefore has an energy close to that of the corresponding state in hydrogen, while the inner electron has an energy equivalent to that of the 1s orbital of a He^+ ion, because the field inside the spherically symmetric charge distribution of the outer orbital is zero. The total energy of such an excited state is therefore close to the sum of the ground-state energy of a He^+ ion and that of the appropriately excited state of the hydrogen atom, which is just the result obtained experimentally and described in the previous paragraph (cf. Fig. 10.1).

Turning our attention now to the ground state of the helium atom, in this case both electrons are associated with the 1s orbital, so the exchange term again vanishes (this time because $u_1 \equiv u_2$) and the effect of the Coulomb term is that each electron screens the other to some extent from the full potential of the doubly charged nucleus; evaluation of the Coulomb integral for the ground state yields a value of 34.0 eV which, when combined with the unperturbed energy of -54.4 eV produces a value of -20.4 eV for the total, in good agreement (considering the limitations of first-order perturbation theory in this context) with the experimental result of -24.8 eV.

Finally, we consider states, other than the ground state, that are not highly excited. We can now no longer assume that the exchange energy is negligible, and the singlet and triplet states are therefore expected to have different energies, as is indeed observed experimentally. Moreover, the effective potential experienced by the electrons is now significantly different from the Coulomb form, so states whose orbitals have the same values of n, but different l, are no longer degenerate. We see in Fig. 10.1 that both these effects are most pronounced for states constructed from an orbital with $n = 2$ along with the 1s orbital, and become progressively smaller as n increases until, for n greater than about five, the states can be considered as 'highly excited' in the sense described above.

10.6 SYSTEMS WITH MORE THAN TWO INDISTINGUISHABLE PARTICLES

The principles discussed in the earlier sections can be readily extended to systems containing more than two particles, but their application to particular cases is often a complex and elaborate process. In this section we shall outline some of the basic ideas to provide an introduction to this topic.

The operators representing the physical properties of a system composed of a number of indistinguishable particles are clearly symmetric with respect to all interchanges of the particle labels, so it follows from the same argument used in the two-particle case that the many-body wave function must always be either symmetric or antisymmetric with respect to the exchange of any pair of labels. Thus, for example, the wave function representing a system of three bosons must obey the conditions

$$\psi(1, 2, 3) = \psi(2, 1, 3) = \psi(3, 2, 1) = \psi(1, 3, 2)$$
$$= \psi(2, 3, 1) = \psi(3, 1, 2) \tag{10.34}$$

while that representing three fermions must satisfy the following

$$\psi(1, 2, 3) = -\psi(2, 1, 3) = -\psi(3, 2, 1) = -\psi(1, 3, 2)$$
$$= \psi(2, 3, 1) = \psi(3, 1, 2) \tag{10.35}$$

We note that in the latter case the wave function is symmetric with respect to a cyclic permutation of the three indices.

The energy eigenvalue equation for non-interacting particles can be separated into as many single-particle equations as there are particles and, assuming these can be solved, the single-particle eigenfunctions $u_1, u_2, \ldots, u_N$ can be formed into products, from which symmetric or antisymmetric linear combinations can be constructed. These can be used as a starting basis in the general case with the interactions included using perturbation theory.

Bosons

It is interesting to note the form of the ground-state eigenfunction of a system of N non-interacting bosons which is

$$u_0(1)u_0(2)\ldots u_0(N) \tag{10.36}$$

where u_0 is the single-particle eigenfunction corresponding to the lowest energy eigenvalue, and all the particles are associated with this function. Such a state plays an important role in determining the properties of gases of bosons at low temperatures when the particles are said to undergo 'Bose–Einstein condensation' into a common ground state. The superfluidity of liquid helium is an example of a phenomenon which can be partly understood using this concept, although a full account requires detailed consideration of the interatomic interactions.

At temperatures above absolute zero the energy eigenfunction of a gas of non-interacting bosons is a symmetric linear combination of products of one-electron functions whose total energy is equal to the internal energy of the gas. This problem can be analysed using statistical methods; the particular form of statistical mechanics in which the exchange symmetry of the wave function is preserved is known as *Bose–Einstein statistics*.

Fermions

In the case of non-interacting fermions, antisymmetric linear combinations have to be formed from the products of single-particle eigenfunctions. The general form of these in the case of a system of N particles where the single-particle eigenfunctions are $u_1, u_2, \ldots, u_N$, is

$$\psi = \frac{1}{(N!)^{1/2}} \begin{vmatrix} u_1(1) & u_2(1) & \cdots & u_N(1) \\ u_1(2) & u_2(2) & \cdots & u_N(2) \\ \vdots & \vdots & & \vdots \\ u_1(N) & u_2(N) & \cdots & u_N(N) \end{vmatrix} \tag{10.37}$$

Such an expression is known as a *Slater determinant*. We can see that this fulfils the required condition of antisymmetry using standard properties of determinants. Thus an exchange of the labels on two particles is equivalent to exchanging two rows of the determinant which leaves its magnitude unchanged, but reverses its sign; also the expansion of an $N \times N$ determinant contains $N!$ terms which leads to the normalization factor of $(N!)^{-1/2}$, assuming the one-electron orbitals are orthonormal. The Pauli exclusion principle also follows from (10.37): if two or more particles are associated with the same single-particle eigenfunction, two or more columns of the determinant are identical and the whole wave function vanishes.

The form of statistical mechanics applying to a gas of non-interacting fermions is known as *Fermi–Dirac statistics*. At zero temperature the wave function has the form (10.37) where the single-particle eigenfunctions are those with smallest eigenvalues, while at higher temperature states with higher energy are involved. We saw an example of a system obeying Fermi–Dirac statistics when we discussed the behaviour of electrons in metals in Chapter 7; another example is a gas of atoms of the isotope ^{3}He whose nucleus has a total spin quantum number of one-half. As the electrons have zero total orbital angular momentum and zero spin in the ground state, the total wave function will be asymmetric with respect to exchange of pairs of hydrogen atoms and such a gas must obey Fermi–Dirac statistics. As a result the low-temperature properties of ^{3}He are quite different from those of normal helium whose nucleus, ^{4}He, has zero spin: whereas the latter exhibits superfluidity below a temperature of about 2 K, the former remains a normal liquid down to temperatures less than 10^{-2} K. At high temperatures, on the other hand, the properties of the two gases are very similar, and this can also be shown to be in agreement with the results of quantum statistical mechanics.

The case of interacting fermions has considerable practical importance because it represents the situation applying in many-electron atoms and molecules as well as in nuclei and in the case of electrons in solids. Exact solutions are rarely possible, but various approximations have been developed. One of the most important of these is known as self-consistent field theory. In this approximation the single-particle eigenfunctions are calculated assuming that each particle is subject to a potential equivalent to the average of its interaction with all the others, along with that due to external forces. We described an elementary application of this principle when we discussed the energies of the highly excited states of helium and, in the more general case of a many-electron atom, the interaction potential is taken to be that from a continuous charge distribution whose magnitude is proportional to the sum of the squared moduli of some approximate set of one-electron eigenfunctions, along with an exchange term analogous to X in (10.32). The resulting eigenvalue equations are then solved to give new one-electron eigenfunctions which in turn are used to construct new potentials. This iterative process is continued until no further significant changes occur, when the system is said to be 'self-consistent'. It can be shown that this procedure is equivalent to using the variational principle to obtain the best possible representation of the many-particle eigenfunction in the form of a single determinant such as (10.37). More accurate results can be obtained if the eigenfunction is represented by a linear combination of determinants, each of which is constructed from a different set of one-electron functions. With the assistance of modern computers, such calculations are quite straightforward for all but the heaviest atoms and for moderately sized molecules. Computations of this kind now form part of the routine armoury of the theoretical chemist and are particularly powerful for the study of short-lived chemical species which can often not be studied in detail experimentally.

10.7 SCATTERING OF IDENTICAL PARTICLES

We close this chapter by considering the problem of scattering where the particles in the incident beam are identical with that constituting the scatterer; an example to which we shall return shortly is the scattering of alpha particles by the nuclei of ^{4}He atoms. We note that in a scattering problem there are no external forces, so we can use the procedure discussed previously for isolated systems to separate the relative motion from that of the centre of mass. Neglecting spin for the present, the energy eigenfunctions can be written as

$$\psi(\mathbf{r}_1, \mathbf{r}_2) = U(\mathbf{R})u(\mathbf{r}) \qquad (10.38)$$

where (using (10.5) and remembering that $m_1 = m_2$) $\mathbf{R} = \frac{1}{2}(\mathbf{r}_1 + \mathbf{r}_2)$ and

$\mathbf{r} = \mathbf{r}_1 - \mathbf{r}_2$. Clearly $U(\mathbf{R})$ is completely symmetric with respect to particle exchange, so the symmetry of the wave function is determined by that of $u(\mathbf{r})$; moreover, particle interchange is equivalent to a reversal of the sign of $\mathbf{r}$ so the exchange symmetry is equivalent to the parity in this case.

In Chapter 9 (cf. Eqs (9.57) to (9.60)) we showed that the energy eigenfunction in a scattering problem could be written in the form

$$V^{-1/2} \left[e^{ikz} + \frac{1}{kr} e^{ikr} f(\theta) \right] \tag{10.39}$$

where

$$f(\theta) = \sum_{l=0}^{\infty} (2l+1) e^{i\delta_l} \sin \delta_l P_l(\cos \theta) \tag{10.40}$$

Equation (10.39) does not have a definite parity so the correct eigenfunction in the case of identical particles must be equal to a linear combination of this expression with a similar one whose sign of $\mathbf{r}$ is reversed. That is,

$$u(\mathbf{r}) = (2V)^{-1/2} \left\{ [e^{ikz} \pm e^{-ikz}] + \frac{1}{r} e^{ikr} [f(\theta) \pm f(\pi - \theta)] \right\} \tag{10.41}$$

where the positive and negative signs apply when $u(\mathbf{r})$ is symmetric and antisymmetric respectively. In the absence of scattering, the wave function is proportional to the first term in square brackets in (10.41), so it follows by an argument similar to that leading to Eqs (9.61) and (9.62) that the differential cross section $\sigma(\theta)$ is given by

$$\sigma(\theta) = |f(\theta) \pm f(\pi - \theta)|^2 \tag{10.42}$$

We know from Chapter 3 that

$$P_l(\cos \theta) = (-1)^l P_l(\cos (\pi - \theta))$$

so it follows from (10.41) and (10.42) that the cross section is made up from partial waves with only even values of l in the symmetric case and with only odd values in the asymmetric case. It also follows from (10.42) that the contributions to the total cross section from those partial waves that have non-zero amplitude are four times what they would be if the particles were distinguishable. Half of this factor of four arises because a recoiling target particle cannot be distinguished from one scattered out of the incident beam, but the remaining factor of two has no such simple physical cause. All these results have been confirmed experimentally by scattering experiments involving spin-zero particles: the scattering of α particles by ^{4}He atoms, for example, departs from the predictions of the Rutherford scattering formula (9.42) in a manner that can be quantitatively accounted for on this basis.

When the particles have non-zero spin, the overall symmetry of the wave function is determined by that of both the spatial and the spin-dependent parts. We shall not consider the general case here, but confine our discussion to the case of spin-half fermions where the total wave function must be antisymmetric. It follows that all partial waves will contribute to the scattering, but for those with even l the spin part must be antisymmetric and therefore correspond to a singlet state with zero total spin, while for those with odd l the spin function must be symmetric, corresponding to a spin-one triplet state. Thus if, for example, the energy of the incident particle were such that the scattering is s-wave in character, we can conclude that scattering will occur only when the particle spins correspond to a singlet state. As, on average, only one quarter of the particle pairs is in this state, the total cross section is reduced by a factor of four to be the same as if the particles were distinguishable. After such a scattering event, therefore, the incident and target particles move off in a state where both the total orbital angular momentum and the total spin are zero. An example of the application of this principle is the scattering of protons by hydrogen nuclei at an energy large enough to ensure that the scattering is predominantly due to the nuclear force rather than the Coulomb interaction, but small enough to ensure that the contribution from partial waves with non-zero values of l is negligible. As will be discussed in the next chapter, measurements of the properties of such pairs of protons can be used to make a valuable comparison of the predictions of quantum mechanics with those of 'hidden variable' theories.

PROBLEMS

10.1 Show that for a many-particle system, subject to no external forces, the total energy and total momentum can always be measured compatibly, but that, if the particles interact, the individual momenta cannot be measured compatibly with the total energy.

10.2 Two particles of masses m_1 and m_2 move in one dimension and are not subject to any external forces. The potential energy of interaction between the particles is given by

$$V = 0 \quad (|x_{12}| \leqslant a); \qquad V = \infty \quad (|x_{12}| > a)$$

where x_{12} represents the particle separation. Obtain expressions for the energy eigenvalues and eigenfunctions of this system if its total momentum is P.

10.3 Repeat the calculation in Prob. 10.2 for the case where the two particles have the same mass m and are (i) indistinguishable spin-zero bosons and (ii) indistinguishable spin-half fermions.

10.4 Show that if $\Psi(\mathbf{r}_1, \mathbf{r}_2, \ldots, \mathbf{r}_N)$ is the wave function representing a system of N indistinguishable particles, then the probability of finding any one of these in the element $d\tau$ around the point $\mathbf{r}$ is given by $P_1(\mathbf{r}) \, d\tau$ where

$$P_1(\mathbf{r}) = \int \ldots \int |\Psi(\mathbf{r}, \mathbf{r}_2, \ldots, \mathbf{r}_N)|^2 \, d\tau_2 \ldots d\tau_N$$

Show that in the case of two non-interacting particles (either bosons or fermions)

$$P_1(\mathbf{r}) = \tfrac{1}{2}(|u_1(\mathbf{r})|^2 + |u_2(\mathbf{r})|^2)$$

where u_1 and u_2 are appropriate single-particle eigenfunctions.

10.5 Two indistinguishable non-interacting spin-half particles move in an infinite-sided one-dimensional potential well. Obtain expressions for the energy eigenvalues and eigenfunctions of the ground and first excited states and use these to calculate $P_1(x)$—as defined in Prob. 10.4—in each case.

Show that in those of the above states where the total spin is zero, there is generally a finite probability of finding the two particles at the same point, but that this probability is zero if the total spin is one.

10.6 Show that if the potential described in Prob. 10.5 has width Na and contains a large number N of non-interacting fermions, the total ground-state energy is approximately equal to $(\pi^2\hbar^2/24ma^2)N$.

10.7 Using the one-dimensional model described in Prob. 10.6, estimate the average energy per particle for (i) a free-electron gas and (ii) a gas of ^{3}He atoms assuming a mean linear density of 4×10^9 particles per metre in each case.

Compare your answer to (ii) with a similar estimate of the ground-state energy per atom of ^{4}He gas. Above about what temperature might you expect these gases to have similar properties?

10.8 A bound system consisting of two neutrons is almost, but not quite, stable. Estimate the energy of the virtual energy level of this system given that the low-energy s-wave scattering cross section of neutrons by neutrons is about 60×10^{-28} m^2.

Hint: cf. Problem 9.9.

ELEVEN

THE CONCEPTUAL PROBLEMS OF QUANTUM MECHANICS

The previous chapters contain a number of examples of the successful application of quantum mechanics to the solution of real physical problems. These of course represent only a small example of the wide range of experimental results, spanning most if not all areas of physics and chemistry and beyond, which have been successfully explained by quantum theory, and so far at least, no quantum-mechanical prediction has been experimentally falsified. Despite these successes, however, the basic conceptual framework of the subject has been considered by many scientists to be unsatisfactory and repeated attempts have been made to reinterpret quantum mechanics, or even to replace it with a different theory whose philosophical and conceptual basis could be considered more acceptable. In the present chapter we shall explain the reasons for this dissatisfaction and outline some of the re-interpretations and alternative approaches which have been devised. Inevitably many of the questions that arise in this area are matters of opinion rather than fact and for this reason some physicists consider that they belong more properly to the realms of philosophy than of physics. However, the conceptual basis of quantum mechanics is so fundamental to our whole understanding of the nature of the physical universe that it should surely be important for physicists to understand at least the nature of the problems involved.†

---

†A more extended and less technical discussion of these ideas can be found in my book *Quantum Physics: Illusion or Reality* (Cambridge University Press, 1986).

11.1 THE CONCEPTUAL PROBLEMS

In this section we outline some of the main areas of difficulty with a view to more detailed discussion later in the chapter.

Determinism

One of the ways in which quantum mechanics differs from classical physics is that the latter is a deterministic theory, which means that the laws are framed in such a way that each event can be seen as a necessary consequence of the theory and the preceding state of affairs. To take a particularly simple example, if we release a massive object in a vacuum, it will certainly fall to the floor and we can calculate the time it will take to do so to a very high accuracy—given the initial position of the object relative to the floor and the acceleration of gravity. In contrast, a given state of a quantum system in general allows us to predict only the relative probabilities of different outcomes. For example, if a spin-half particle known to be in an eigenstate of S_z enters a Stern–Gerlach apparatus oriented to measure S_x, we know that a positive and a negative result are equally likely, but which actually occurs is completely unpredictable.

It is important to realize that, although quantum physics is generally applied to the microscopic world of atoms and the like, indeterminism can also affect macroscopic events. Thus, if we have detectors in each channel of the Stern–Gerlach experiment described above, what is uncertain is which of these will 'click' and record the passage of an atom. This counter click is a perfectly macroscopic event and could be used to trigger any other everyday action. For example, we could decide that if one counter clicked, we would go to work, while if the other did so we would go back to bed. The possibly earth-shattering consequences of this decision are then completely indeterminate and cannot be predicted in advance by any observation we might make on the system before the experiment is performed.

This unpredictability, or indeterminism, of quantum mechanics can be traced back to the measurement postulate (Postulate 4.4 in Chapter 4) where the procedure for predicting the relative probabilities of different outcomes of a measurement is defined. It is important to note that it is here, in the measurement, that the uncertainty arises. In contrast, the evolution of the wave function between measurements is perfectly deterministic. We mean by this that, if we know the wave function at any time, we can in principle use the time-dependent Schrödinger equation to calculate its form at any future time, and the fact that this is a linear equation means that there is always a unique outcome to this process. Referring again to the Stern–Gerlach example, the wave function of a particle that was initially in an S_z eigenstate will emerge from the final (S_x) measurement as a wave packet that is split into two parts, corresponding to the two paths that would have been followed by particles initially in eigenstates of $\hat{S}_x$. It is only when we 'measure' the result of the experiment by allowing the particle to interact with a counter that the indeterminacy occurs. At this stage, only one of the two counters actually fires and the wave function becomes that corresponding to the appropriate eigenfunction of the operator corresponding to the measurement; this process is sometimes known as the 'reduction' or 'collapse' of the wave function. We are forced to conclude that there are two different rules for calculating time dependence in quantum mechanics: the time-dependent Schrödinger equation which controls the deterministic evolution of the wave function, and the measurement postulate which produces the random indeterministic results of 'actual' measurements. Defining the boundary between these two processes constitutes the 'measurement problem' in quantum mechanics to which we shall return in more detail later in this chapter.

11.2 HIDDEN-VARIABLE THEORIES

Many of the conceptual problems of quantum mechanics would be resolved if it could be shown that the predictions of the theory were statistical, in the sense that they described the relative probabilities of occurrences which are in fact determined by properties that cannot themselves be observed directly. An analogy is sometimes drawn with classical statistical mechanics where different possible events are ascribed particular probabilities, even though a detailed examination of the behaviour of the component atoms would show that these were following perfectly deterministic laws. If a similar deterministic substructure could be found to underlie quantum mechanical indeterminism, this would have many attractions. In the quantum case, however, we do not know what the substructure is, or even if it exists at all. For this reason any such quantities that may be postulated to underlie quantum behaviour and cannot be directly observed are known as 'hidden variables', and theories based on such ideas are called 'hidden-variable theories'.

One necessary property of any hidden-variable theory is that it must reproduce the results of quantum mechanics in every case where these have been confirmed. Of course, if an as yet unperformed experiment could be devised in which the predictions of the new theory differed from those of quantum mechanics, then we should be in a position to make an experimental test to decide which of the two is correct. To date, all such tests have come down on the side of quantum mechanics, but if the opposite is ever found to be true, it will constitute one of the most important discoveries of modern physics.

De Broglie–Bohm theory

The earliest example of a hidden-variable theory was that proposed by de Broglie in 1927 as a direct consequence of his postulate of the existence of matter waves. This initial work has been developed and extended by a number of other workers since—notably David Bohm, who rediscovered de Broglie's ideas in the 1950s. For this reason it has become known as the de Broglie–Bohm theory.

The basic idea behind de Broglie–Bohm theory is that atomic particles such as electrons always possess a real position and velocity. The matter wave also exists and acts to guide the motion of the particles in such a way that their statistical properties are just those predicted by quantum mechanics. We can see how this works if we start from the Schrödinger equation governing the motion of a particle of mass m in a potential $V(\mathbf{r})$:

$$i\hbar\,\partial\psi/\partial t = -(\hbar^2/2m)\nabla^2\psi + V\psi \tag{11.1}$$

We now define quantities R and S as real functions of $\mathbf{r}$ such that

$$\psi = R\exp(iS/\hbar) \tag{11.2}$$

If we substitute from (11.2) into (11.1) we get

$$i\hbar(\partial R/\partial t)\exp(iS/\hbar) - R(\partial S/\partial t)\exp(iS/\hbar)$$
$$= -(\hbar^2/2m)\,[\nabla^2 R + 2i(\nabla R)\cdot(\nabla S)/\hbar + iR\nabla^2 S/\hbar$$
$$- R(\nabla S)^2/\hbar^2]\exp(iS/\hbar) + VR\exp(iS/\hbar) \tag{11.3}$$

We now cancel the common factor $\exp(iS/\hbar)$ and separate the real and imaginary parts to get two equations:

$$\left.\begin{array}{l} \partial S/\partial t + (1/2m)(\nabla S)^2 + V - (\hbar^2/2m)(\nabla^2 R/R) = 0 \\ \partial R/\partial t + (\nabla R)\cdot(\nabla S/m) + R\nabla^2 S/2m = 0 \end{array}\right\} \tag{11.4}$$

If we multiply the second of (11.4) by R and rearrange we get

$$\partial R^2/\partial t + \nabla\cdot(R^2\nabla S/m) = 0 \tag{11.5}$$

We can also rewrite the first of (11.4) as

$$\partial S/\partial t + (1/2m)(\nabla S)^2 + V - Q = 0 \tag{11.6}$$

where Q is defined by

$$Q = -(\hbar^2/2m)\nabla^2 R/R \tag{11.7}$$

All we have done so far is to rewrite the Schrödinger equation in terms of the new functions R and S. However, we can now see that this recasting leads to the physical ideas underlying de Broglie–Bohm theory. First we note that R^2 equals $|\psi|^2$, which is the probability density of finding a particle in the vicinity of $\mathbf{r}$. If we make the further assumption that a particle at the position $\mathbf{r}$ has a velocity $\mathbf{v}$ where $\mathbf{v} = \nabla S/m$, then (11.5) is simply the equation of continuity, which states that the rate of increase (or decrease) of the probability density in any element of volume must equal the average net flow of particles into (or out of) it. Given this expression for the particle velocity, the second and third terms of (11.6) are now just the kinetic and potential energies of the particle. The fourth term, Q, however, has no classical analogue; according to the de Broglie–Bohm theory, it is an additional potential known as the 'quantum potential'. It is a fundamental feature of the model that the difference between classical and quantum behaviour arises solely from this additional term. We can see how this comes about in the particular case of a conservative system where V is independent of time. It then follows from the separation of variables in the time-dependent Schrödinger equation (cf. Sec 3.1) that $\partial S/\partial t$ equals $-E$, the total energy of the system. Equation (11.6) is then equivalent to the classical conservation of energy equation, with the potential modified by the addition of the quantum potential. It can be shown that in the more general case where the 'real' potential, V, is time-dependent, (11.7) reduces to what is known as the Hamilton–Jacobi formulation of Newtonian mechanics.

It is interesting to see how the de Broglie–Bohm theory accounts for quantum behaviour in one or two particular cases. First, consider the case where V is zero or uniform throughout space. We know that the solution of the Schrödinger equation in this case is a plane wave of constant amplitude (see Chapters 2 and 3), so it follows from (11.7) that the quantum potential is zero everywhere. The particle therefore feels no force and moves with constant velocity, just as we would expect. If, however, the particle passes through one or more slits in a screen, there will be regions in the vicinity of the diffraction minima where the amplitude, R, is very small. This leads to the quantum potential, Q, being very large and negative (cf. 11.7), which in turn results in strong repulsive forces which keep the particles out of the region. Another example is barrier penetration by the tunnel effect (cf. Chapter 2); the wave function is now a travelling wave packet, which gives rise to a fluctuating quantum potential; this in turn can cause an occasional lowering of the total potential barrier and hence the transmission of a particle from time to time.

In recent years, the solutions to equations (11.5) and (11.6) have been computed for a number of different potentials. The actual trajectories followed by the particles (according to the theory) have been obtained, and when the statistical behaviour of a number of particles is calculated, this is always found to agree with the predictions of quantum mechanics—which is not surprising as the whole theory is constructed on this basis. An example of the trajectories associated with diffraction by a double slit is shown in Fig. 11.1. The statistical distribution of the particles arriving at the screen is just what is predicted by quantum mechanics and observed experimentally, yet the particles follow clearly defined paths, just as they would in classical mechanics. Moreover, the model makes some additional statements: for example, that all particles observed in the top half of the screen have passed through the upper slit and vice versa. However, these predictions are completely untestable experimentally because any modification of the

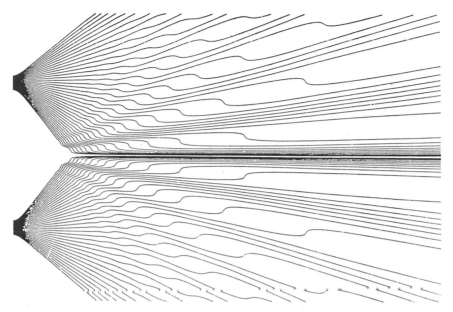

Figure 11.1 The particle trajectories calculated for double-slit diffraction using the de Broglie–Bohm hidden-variable theory. The trajectories cluster around the diffraction maxima in the expected way. (Reproduced by permission from C. Philippides, C. Dewdney and B. J. Hiley, *Nuovo Cimento* B **52** 15–29 (1979). Copyright Società Italiana di Fisica.)

apparatus that would seem to allow us to check this point would result in a change of the wave function and hence the quantum potential to such an extent that the interference pattern would disappear: the Heisenberg uncertainty principle is applicable to de Broglie–Bohm theory just as it is to quantum mechanics.

De Broglie–Bohm theory has been extended to include spin, and the results of Stern–Gerlach experiments of the type described earlier in this chapter have been reproduced. The model is so successful, in fact, that the reader could be forgiven for wondering why it is not taken to be the orthodox interpretation of quantum mechanics and why it is not taught from the start, so avoiding many of the conceptual difficulties associated with the conventional approach.

There are a number of reasons why de Broglie–Bohm theory has not gained universal acceptance. These are largely associated with the nature of the 'quantum potential'. Potentials in physics, both classical and quantum, arise as a result of physical interactions between objects. For example, the potential experienced by an electron in a hydrogen atom is due to the electromagnetic interaction between the negatively charged electron and the positive proton. But the quantum potential has no obvious physical basis: it arises purely out of the mathematics of the Schrödinger equation. A more philosophical objection to de Broglie–Bohm theory is that it makes unnecessary postulates: the particle trajectories can never be observed more accurately than the uncertainty principle allows, so the postulate that they exist, along with the equally unnecessary quantum potential, is seen as an unnecessary extravagance. The supporters of the theory, on the other hand, believe that this extravagance is preferable to the conventional approach which abjures any realistic description of the behaviour of quantum objects. But the most important reason why de Broglie–Bohm theory is not generally accepted is that it is a 'non-local' theory. By this, we mean that the particle may be influenced not only by the potential at the point where the particle is, but also by its values at other points in space. This may not be clear in the formulation set out above, because we have restricted our consideration to the special case of a single particle moving in a potential $V(\mathbf{r})$. If we extend the argument to the case of two interacting particles, then it follows immediately from the discussion in Chapter 10 that the wave function and hence the quantum potential are functions of the coordinates of both particles, and therefore do not exist in real space but in an abstract six-dimensional 'configuration space' spanned by the six coordinates. An appropriate change in the 'real' potential $V(\mathbf{r}_1, \mathbf{r}_2)$ in the vicinity of one particle produces an immediate change in the joint wave function and hence the quantum potential affecting the other particle—even though the latter may be a large distance from the former. If such a non-local quantum potential is accepted, then indeed all the results of quantum theory (at least in the non-relativistic regime to which this book is confined) can be reproduced by de Broglie–Bohm theory. However, the majority of quantum physicists are unwilling to accept a model based on such 'action at a distance' and prefer the more conventional approach which avoids the problem of non-locality by excluding any realistic description of the properties of quantum systems beyond those that can be directly observed.

If de Broglie–Bohm theory is not acceptable because of its non-locality, might there be another form of hidden-variable theory that could 'explain' quantum mechanics without having this disadvantage? In the 1950s this question very much interested the quantum physicist John Bell, who believed that de Broglie–Bohm theory has been 'scandalously neglected'. He was able to show that no local hidden-variable theory can ever reproduce all the results of quantum mechanics. Because this result is so important and has been so influential over the recent development of thought in this area, we devote the next section to a reasonably detailed discussion of it.

11.3 NON-LOCALITY

The non-local implications of quantum mechanics were first discussed by A. Einstein, B. Podolski and N. Rosen in a paper published in 1925, and work in this area is often referred to as the 'EPR' problem after the initials of their surnames. The problem, though not its solution, was greatly clarified by an illustrative example proposed by David Bohm in the 1960s. This consists of a pair of spin-half particles, each of which is known to have zero orbital angular momentum and whose total spin is also known to be zero. Such a system can be created if a beam of low-energy protons undergoes *s*-wave scattering from hydrogen gas; in this case, as we saw at the end of Chapter 10, the incident and target protons move apart in a state such that the total orbital angular momentum and the total spin of the pair are both zero. The particles are allowed to move apart until they are widely separated and then we measure (say) the z component of spin of one particle (S_{z1}) and follow this by measuring the same component of the other (S_{z2}). Because the total spin is zero, we would expect equal and opposite results $(\pm \frac{1}{2}\hbar)$, and this is indeed what is observed. However, this result actually implies some kind of non-local interaction between the two measurements. Remember that in quantum mechanics a measurement results in a transformation of the wave function of a system from what it was previously into an eigenstate of the measurement operator. Thus the measurement of S_{z1} will result in the transformation of the whole wave function into one that is an eigenstate of both S_{z1} and S_{z2}. This means that, once the first measurement has been made, the results of the second are completely predictable. Indeed the second measurement is unnecessary: we know exactly what its consequences will be as soon as the first measurement has been made, *even though the two particles are widely separated and there is no known interaction between them.* It is important to note that this differs from the equivalent situation in classical mechanics. It is true that classically, if an object with zero angular momentum splits into two parts and separates, then the measurement of the angular momentum of one part tells us that that of the other will be equal and opposite. However, in the classical case, the act of measurement has no effect on the system, while the opposite is true in the quantum case where the whole of measurement theory depends on the wave function of the system being influenced by the act of measurement. In a hidden-variable theory, the properties of the system are controlled by the hidden variables—the quantum potential in the de Broglie–Bohm case—so the above argument makes it unlikely that any local hidden-variable theory could reproduce these results.

Although the above argument strongly suggests that non-locality is an essential feature of such systems, it does not actually prove that the results could not be reproduced by some local hidden-variable theory. Bell's theorem provides such a proof. Before we can proceed to this, we will need the quantitative predictions of quantum mechanics for a measurement of the component of the second spin in a direction at an angle ϕ to the z axis $(S_{\phi 2})$ following a measurement of the z component of spin of the first particle (S_{z1}). We shall assume that all angular momenta are expressed in units of $\frac{1}{2}\hbar$ so that the result of any such measurement is either $+1$ or -1. In the case where the result of the first measurement is positive, we conclude that S_{z2} is consequently negative and that the spin part of the wave function of this particle can be represented by the column vector $\begin{bmatrix} 0 \\ 1 \end{bmatrix}$, as in Table 6.1.

The eigenvectors of $S_{\phi2}$ are also given in Chapter 6 as $\begin{bmatrix} \cos(\phi/2) \\ \sin(\phi/2) \end{bmatrix}$ and $\begin{bmatrix} -\sin(\phi/2) \\ \cos(\phi/2) \end{bmatrix}$ when the eigenvalues are positive and negative respectively. We can now use the measurement postulate to expand the initial wave vector as a linear combination of the eigenvectors:

$$\begin{bmatrix} 0 \\ 1 \end{bmatrix} = \sin(\phi/2) \begin{bmatrix} \cos(\phi/2) \\ \sin(\phi/2) \end{bmatrix} + \cos(\phi/2) \begin{bmatrix} -\sin(\phi/2) \\ \cos(\phi/2) \end{bmatrix} \quad (11.8)$$

so the probability of the result of the second measurement being also positive is $P_{++}(\phi)$ where $P_{++}(\phi) = \sin^2(\phi/2)$ and that of it being negative is $P_{+-}(\phi) = \cos^2(\phi/2)$. Similar arguments lead to expressions for the similarly defined probabilities $P_{-+}(\phi)$ and $P_{--}(\phi)$, and we have

$$\left. \begin{aligned} P_{++}(\phi) &= \sin^2(\phi/2) \\ P_{+-}(\phi) &= \cos^2(\phi/2) \\ P_{-+}(\phi) &= \cos^2(\phi/2) \\ P_{--}(\phi) &= \sin^2(\phi/2) \end{aligned} \right\} \quad (11.9)$$

We shall later wish to refer to a correlation coefficient $C(\phi)$ which is defined as the average value of the product $S_{z1}S_{\phi2}$ when measurements have been performed on a large number of particle pairs. It follows directly from this definition and those of the above probabilities that

$$\begin{aligned} C(\phi) &= \tfrac{1}{2}[P_{++}(\phi) + P_{--}(\phi) - P_{+-}(\phi) - P_{-+}(\phi)] \\ &= \sin^2(\phi/2) - \cos^2(\phi/2) \\ &= -\cos\phi \end{aligned} \quad (11.10)$$

Before proceeding to the general arguments leading to Bell's theorem, it is instructive to consider a particular theory in which we try to adapt the classical theory of angular momentum to the quantum situation. Because this is to be a local theory it is necessarily different from the de Broglie–Bohm theory which, as we have seen, will always reproduce the results of quantum mechanics. We first postulate that all the components of angular momentum of a particle always have definite, if unknown, values. It follows that, as all the components of the total combined spin of the two particles are zero, the 'real spin' vectors associated with the two individual particles are equal and opposite and remain so as the particles separate. In quantum mechanics only one component of spin can be measured without affecting the other two and we shall assume that this property holds in our model. A further assumption has to be made concerning the interaction between the 'real spin' and the measuring apparatus if the only possible results of the measurements are to be plus or minus one (in units of $\frac{1}{2}\hbar$): we postulate that if any spin component is 'really' positive (or negative) then a measurement will always yield the result $+1$ (or -1) whatever the actual magnitude of the component.

It follows directly from the above that, if S_{z1} is measured as $+1$, then the z component of the second spin must be negative and the 'real' spin vector of the latter must lie somewhere on a hemisphere whose symmetry axis is the negative z axis (see Fig. 11.2). Moreover if the component $S_{\phi 2}$ is positive also, the corresponding true spin must lie somewhere in a hemisphere whose symmetry axis makes an angle ϕ with z. Given that the absolute orientation in space of the two spin vectors is random—provided of course that they are always equal and opposite to each other—the probability, $P'_{++}(\phi)$, that both components are positive is proportional to the volume of overlap of the two hemispheres. We see from Fig. 11.2 that this is in turn proportional to ϕ. As we know that when $\phi = \pi$ the result must be a certainty, that is, $P'_{++}(\pi) = 1$, the constant of proportionality must be equal to $1/\pi$. The other probabilities can be estimated in a similar manner so we have

$$\left. \begin{aligned} P'_{++}(\phi) &= \frac{\phi}{\pi} \\[2mm] P'_{+-}(\phi) &= 1 - \frac{\phi}{\pi} \\[2mm] P'_{-+}(\phi) &= 1 - \frac{\phi}{\pi} \\[2mm] P'_{--}(\phi) &= \frac{\phi}{\pi} \end{aligned} \right\} \qquad (11.11)$$

and hence the correlation coefficient $C'(\phi)$, measuring the average of the product of the two measured components is given by (cf. 11.10)

$$C'(\phi) = \frac{2\phi}{\pi} - 1 \qquad (11.12)$$

The results of our 'real spin' theory could be identical with those of quantum mechanics only if the quantities defined in (11.11) and (11.12) were the same as the corresponding expressions in (11.9) and (11.10). Comparison of these two sets of equations shows that, although the theories agree when the two sets of measuring apparatus are either parallel or perpendicular (i.e., when $\phi = 0$, $\pi/2$, or π) there are considerable disagreements at other orientations, and this is illustrated in Fig. 11.3 which shows graphs of $C(\phi)$ and $C'(\phi)$ as functions of ϕ. Thus we have shown that the predictions of quantum

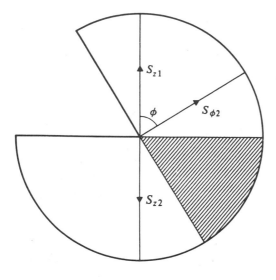

Figure 11.2 According to the 'real spin' hidden-variable theory, the number of particle pairs where the z component of the spin of the first particle and that of the second particle in the direction defined by ϕ are both positive is proportional to the shaded area.

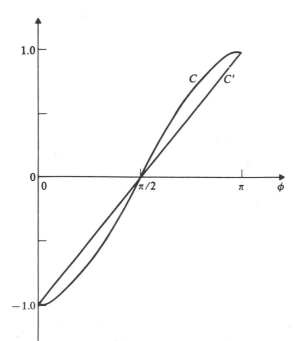

Figure 11.3 The spin correlation function calculated by quantum mechanics (C) and using a 'real spin' hidden-variable theory (C').

mechanics do not agree with those of a hidden-variable theory based on the apparently reasonable assumption that the particles possess real angular momentum vectors that interact with Stern–Gerlach apparatuses in the manner described above. We could try to modifiy our model in some way so as to improve the agreement, but, as we shall see, such an attempt would be bound to fail because any local hidden-variable theory must produce results which disagree with quantum mechanics.

Bell's Theorem

We now proceed with this general proof and first consider what are the minimum requirements of a local deterministic hidden-variable theory. These are that, after the particles separate, the system must possess some property that determines in advance the result of performing a Stern–Gerlach measurement of any component of the spin of either particle. In the previous example this property consisted of the equal and opposite 'real spins', but in general we need not identify the hidden variables with the parameters of any particular physical model. Accordingly we represent the result of measuring the z component of spin of the first particle by $S_{z1}(\lambda)$ where, as before, S_{z1} has the value plus one or minus one (in units of $\frac{1}{2}\hbar$) and λ represents a hidden variable. (There may of course be more than one hidden variable—for example there were three 'real spin' components in the model used earlier—but the argument can be readily generalized to include this case.) Similarly, the result of a measurement of the spin component of the second particle in a direction at an angle ϕ to the z axis is also determined by λ and accordingly written as $S_{\phi2}(\lambda)$. Such a hidden-variable theory is deterministic, because the values of S_{z1} and $S_{\phi2}$ are determined by the value of λ, and also local, because the result of each measurement is independent of the direction chosen for the other. Each pair of particles possesses a particular value of λ, and we define a quantity $p(\lambda)$ so that the probability of a pair being produced with a value of λ between λ and $\lambda + d\lambda$ is $p(\lambda)\,d\lambda$; we also assume this to be normalized so that

$$\int p(\lambda)\,d\lambda = 1 \qquad (11.13)$$

We now consider an experiment in which the quantities S_{z1} and $S_{\phi2}$ are measured on a large number of particle pairs. According to our hidden-variable theory, the average of the values of the products $S_{z1}\,S_{\phi2}$ will be given by $C''(\phi)$ where

$$C''(\phi) = \int S_{z1}(\lambda)S_{\phi2}(\lambda)p(\lambda)\,d\lambda \qquad (11.14)$$

Clearly $C''(\phi)$ is a similar correlation coefficient to that previously denoted by $C(\phi)$ and $C'(\phi)$. If we now consider a separate set of measurements in which the first apparatus is as before, but the second apparatus is oriented to measure the component at an angle θ to the z axis, we obtain a similar expression for the quantity $C''(\theta)$ and hence

$$C''(\phi) - C''(\theta) = \int [S_{z1}(\lambda)S_{\phi2}(\lambda) - S_{z1}(\lambda)S_{02}(\lambda)]p(\lambda)\,d\lambda \quad (11.15)$$

Now we know that if we measure the same component of both spins, we shall always obtain equal and opposite results. That is

$$S_{\theta 2}(\lambda) = -S_{\theta 1}(\lambda) \\ S_{\phi 2}(\lambda) = -S_{\phi 1}(\lambda) \quad\Big\} \tag{11.16}$$

Substituting from (11.16) into (11.15) we get

$$C''(\phi) - C''(\theta) = -\int S_{z1}(\lambda)[S_{\phi 1}(\lambda) - S_{\theta 1}(\lambda)]p(\lambda)\, d\lambda$$
$$= -\int S_{z1}(\lambda)S_{\phi 1}(\lambda)[1 - S_{\phi 1}(\lambda)S_{\theta 1}(\lambda)]p(\lambda)\, d\lambda \tag{11.17}$$

where the last step follows because $S_{\phi 1}(\lambda) = \pm 1$ and therefore $[S_{\phi 1}(\lambda)]^2 = 1$. If we now take the absolute value of both sides of (11.17) we get

$$|C''(\phi) - C''(\theta)| = |\int S_{z1}(\lambda)S_{\phi 1}(\lambda)[1 - S_{\phi 1}(\lambda)S_{\theta 1}(\lambda)]p(\lambda)\, d\lambda|$$
$$\leqslant \int |S_{z1}(\lambda)S_{\phi 1}(\lambda)[1 - S_{\phi 1}(\lambda)S_{\theta 1}(\lambda)]p(\lambda)|\, d\lambda \tag{11.18}$$

because the absolute value of an integral is always less than or equal to the integral of the absolute value of the integrand. Now neither $p(\lambda)$ nor the term in square brackets in (11.18) can ever be negative, and moreover $|S_{z1}(\lambda)S_{\phi 1}(\lambda)| = 1$ because each of these quantities equals ± 1, so (11.18) can be written as

$$|C''(\phi) - C''(\theta)| \leqslant \int [1 - S_{\phi 1}(\lambda)S_{\theta 1}(\lambda)]p(\lambda)\, d\lambda$$
$$= 1 + \int S_{\phi 1}(\lambda)S_{\theta 2}(\lambda)p(\lambda)\, d\lambda \tag{11.19}$$

using (11.13). We shall confine our discussion to cases where the z axis and the directions defined by θ and ϕ are in the same plane; moreover, the averages obtained should depend only on the relative orientations of the three measurements and not on the absolute direction of the z axis (this last assumption can in fact be tested by a separate set of experiments). It follows that the integral in (11.19) measures the correlation between two spin components that are at an angle of $(\theta - \phi)$ with each other and is therefore equal to $C''(\theta - \phi)$. We can therefore write (11.19) as

$$|C''(\phi) - C''(\theta)| - C''(\theta - \phi) \leqslant 1 \tag{11.20}$$

The inequality (11.20) is known as *Bell's theorem* and, as we have seen, it represents a necessary consequence of any local deterministic hidden-variable theory. We now wish to see whether or not it is consistent with the predictions of quantum mechanics. To do this it is most convenient to consider the special case where θ equals 2ϕ. It then follows from (11.10) that the quantum-mechanical expressions for the corresponding correlation coefficients are

$$C(\phi) = -\cos\phi \\ \tag{11.21}$$
$$\text{and} \qquad C(2\phi) = -\cos 2\phi \quad\Big\}$$

Comparison of (11.20) and (11.21) shows that the predictions of quantum

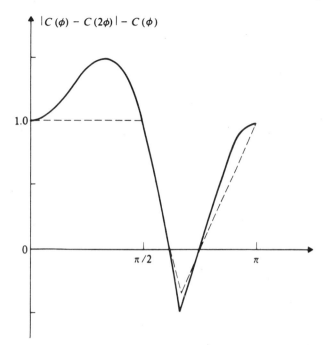

Figure 11.4 The continuous line shows the quantity $|C(\phi) - C(2\phi)| - C(\phi)$ calculated by quantum mechanics, while the broken line shows the equivalent quantity—that is, $|C'(\phi) - C'(2\phi)| - C'(\phi)$—calculated using the 'real spin' hidden-variable theory. According to Bell's theorem this quantity must always be $\leqslant 1$ for any local hidden-variable theory.

mechanics could be consistent with those of a hidden-variable theory only if $|\cos 2\phi - \cos \phi| + \cos \phi \leqslant 1$. Figure 11.4 shows a graph of this quantity as a function of ϕ and we see that, although Bell's theorem is obeyed in the region $\pi/2 \leqslant \phi \leqslant \pi$, there is considerable disagreement with quantum mechanics in the range $0 \leqslant \phi \leqslant \pi/2$ which reaches a maximum at $\phi = \pi/3$ where the function equals 1.5. Figure 11.4 also shows the equivalent quantity calculated on the basis of the 'real spin' hidden-variable theory discussed above; as expected this is consistent with Bell's theorem and disagrees with quantum mechanics, although it is interesting to note that this discrepancy is least in the region where the predictions of quantum mechanics are consistent with Bell's theorem.

We are therefore forced to conclude that no local deterministic hidden-variable theory is capable of reproducing the results of quantum mechanics for experiments such as those described. As quantum mechanics has been so well verified over a wide range of phenomena, it might be thought that any further experimental test of the discrepancy between the predictions and those of hidden-variable theories would be unnecessary. However, it turns out that, because of the particular nature of the correlation discussed above, none of the experiments that had been performed before the formulation of

Bell's theorem in 1969 provided a direct test of this point. In fact the form of Bell's theorem set out in (11.20) cannot be tested directly as it assumes that all particle pairs have been detected while, in any experiment conducted with real detectors, some particles are always missed. However, other versions of Bell's theorem have been derived that are not subject to this criticism. Moreover, the proofs have been extended to include hidden-variable theories that are not perfectly deterministic and include an element of randomness in determining their predictions.

Although Bell's theorem clearly demonstrates that no hidden-variable theory can reproduce the results of quantum mechanics, the implications of non-locality are so profound that some thought that it was more likely that quantum mechanics would prove incorrect in such situations than that the predictions of Bell's theorem would be upheld, and over the past twenty years or so a number of experiments have been performed to test this point. It turns out that, for technical reasons, these are difficult to perform on spin-half particles and most have been carried out using pairs of photons. The polarization of a photon follows quantum rules that are practically identical to those governing the spin of a spin-half particle: photon polarization can be measured as being directed parallel or perpendicular to some direction in space and such measurements follow rules that are essentially identical to those applying to the measurement of spin as 'up' or 'down'. By common consent, the best experiments to date were performed by Alain Aspect in France in the early 1980s and Fig. 11.5 shows the essentials

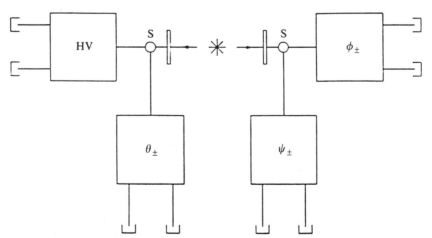

Figure 11.5 The Aspect experiment to compare the predictions of quantum mechanics with those of Bell's theorem for pairs of polarized photons. The photons can be switched into either channel on each side, so that measurements can be made with four different pairs of polarizations. The switches marked S are changed at a rate faster than the time required to send a signal from one side of the apparatus to the other at the speed of light. (Reproduced with permission from *Quantum Physics: Illusion or Reality?* by A. I. M. Rae, Cambridge University Press, 1986.)

of his apparatus. Pairs of photons are emitted from the centre of the apparatus in a quantum state such that the polarizations of the members of each pair are always perpendicular (just as the spins of the atom pairs discussed earlier are opposite). Each photon is then directed into one of two polarizers that determines whether its spin is parallel or perpendicular to some direction. Which polarizer a particular photon enters is determined by a switch that operates at a frequency of around 10^8 Hz. The polarizers are oriented so that, as in the diagram, one of the left-hand polarizers is set to measure polarization as either horizontal or vertical (H/V) and the other as either parallel or perpendicular to a direction at an angle θ to the horizontal, while those on the right-hand side are set at ϕ and ψ to the horizontal respectively. After a number of photon pairs have passed through the apparatus we can determine the number which have been measured as, say, vertical on the left-hand side and parallel to ϕ on the right, which we call $n(V, \phi^+)$. Other pairs are similarly denoted. Separate runs are performed in which the left-hand photon is detected with the polarizers removed on that side and the right-hand photon is detected with its polarization parallel to θ. The number detected in this way is denoted as $n(\theta^+)$. The extended version of Bell's theorem applicable to this situation is

$$n(V, \phi^+) - n(V, \psi^+) + n(\theta^+, \phi^+) + n(\theta^+, \psi^+) \leqslant n(\theta^+) + n(\phi^+) \quad (11.22)$$

It can be shown that the maximum disagreement with the predictions of quantum mechanics for this set-up occurs when the angles are $\theta = 45°$, $\phi = 67\frac{1}{2}°$ and $\psi = 22\frac{1}{2}°$, when the left-hand side should be larger than the right by $0.112N$, where N is the total number of photon pairs recorded in each run. The experimental result in this configuration was $0.101N$, which agrees with quantum prediction within experimental error, but is completely inconsistent with Bell's theorem.

The Aspect experiment has another property that is of considerable significance. This is that, because the switches operate at such a high rate, the time between switchings is considerably shorter than the time needed for light to travel from one side of the apparatus to the other. Thus, if we were to imagine that the correlations were established as a result of some unknown physical interaction between one side of the apparatus and the other, then this would have to be propagated faster than light. Overall then, the Aspect experiment seems to provide conclusive proof that a correct hidden-variable theory must be non-local in nature and the non-local influence must be propagated through space instantaneously, or at least at superluminal speed. This is certainly the conclusion drawn by nearly all workers in the field, but it should be noted that the argument is not quite rigorous. This is because the proof of the extended Bell's theorem relies on several additional assumptions, among which is the apparently reasonable one that polarizers may attenuate, but can never amplify, light. T. W. Marshall and others have pointed out that, because the efficiency of photodetectors is quite low, it is possible to postulate a breach of this 'no enhancement' postulate without implying any overall increase in the energy of the system, and so to produce a local hidden-variable theory that is compatible with the experimental results obtained so far. However, the majority view is that, although this loophole does exist, it is very small and will eventually be closed when high-efficiency photodetectors are developed.

If we accept that the Aspect experiment rules out local hidden-variable theories, and if we are not convinced that non-local theories such as that of de Broglie–Bohm are correct, then what is left? The obvious recourse is to return to quantum mechanics and look again at the standard interpretation. An important feature of this is that properties that cannot in principle be measured in a particular experimental situation do not have any real existence. Thus, if we pass a spin-half particle through a Stern–Gerlach apparatus oriented to measure S_z, then the other spin components are not just unmeasurable but have no reality. The traditional, 'Copenhagen', interpretation of quantum mechanics, propounded by Niels Bohr in particular, assigns physical properties not to individual quantum systems on their own, but only to quantum systems in conjunction with measuring apparatus; the only physical properties that are real in this situation are those that can be measured by the apparatus. The proof of Bell's theorem then fails because the unmeasured spin components do not have values that are functions of the hidden variables: they just do not exist. Clearly, however, such an interpretation of quantum mechanics implies that we know what a measurement is, and this brings us back to the measurement problem.

11.4 THEORIES OF MEASUREMENT

Probably the most difficult and controversial conceptual problem in quantum mechanics concerns the nature and meaning of the quantum theory of measurement described in Chapter 4. The relevant theory is contained in Postulates 4.2, 4.3, and 4.4 which state that the measurement of a physical quantity always produces a result equal to one of the eigenvalues (q_n) of the operator ($\hat{Q}$) representing that quantity, that the wave function immediately after the measurement is the same as the corresponding eigenfunction (ϕ_n), and that if the wave function is ψ before the measurement, the probability of obtaining the result q_n is equal to $|c_n|^2$ where $\psi = \Sigma_n c_n \phi_n$. The effect of the measurement is therefore to cause the wave function to be changed (or *reduced*) from ψ to ϕ_n. We can represent this process by

$$\psi \xrightarrow[\substack{\text{measurement}\\\text{giving}\\\text{result } q_n}]{} \phi_n \tag{11.23}$$

In order to consider this question further, we must look more closely at what actually constitutes a measurement. To do this we consider once again the example of the measurement of a component of spin of a spin-half particle using a Stern–Gerlach apparatus, noting that it is not necessary now to consider two correlated particles. If the component to be measured is S_z, whose eigenvalues are $\pm\frac{1}{2}\hbar$ and whose eigenvectors are represented by α_z and β_z, and if the initial spin state of the particle is α_x, Eq. (11.23) becomes in this case

$$
\left.
\begin{aligned}
\alpha_x &= \frac{1}{\sqrt{2}}(\alpha_z + \beta_z) \xrightarrow[\substack{\text{measurement}\\\text{yielding}\\ S_z = \frac{1}{2}\hbar}]{} \alpha_z \\[2em]
\alpha_x &= \frac{1}{\sqrt{2}}(\alpha_z + \beta_z) \xrightarrow[\substack{\text{measurement}\\\text{yielding}\\ S_z = -\frac{1}{2}\hbar}]{} \beta_z
\end{aligned}
\right\}
\qquad (11.24)
$$

An important point to note is that this measurement, and the subsequent reduction of the wave function, are not achieved simply by passing the particles through an appropriately oriented Stern–Gerlach magnet. This follows from a consideration of the arrangement illustrated in Fig. 11.6 where particles with a known S_x are passed through such a magnet oriented to 'measure' S_z and then directed back into a common path so that it is impossible to tell through which channel any particular particle passed. Thus no information has actually been obtained about the value of S_z for any particle and indeed, the wave function is not reduced, its spin part remaining unchanged as α_x after the 'measurement'. It follows that in order to make a successful measurement of S_z, some detecting device or counter must be introduced to record through which channel the particle passed. It is the presence of such recording apparatus which apparently causes the reduction of the wave function described by (11.23) and (11.24).

The description of the quantum theory of measurement in the previous paragraph would be perfectly correct and sufficient, provided we could treat the recording equipment as a separate piece of apparatus obeying the laws

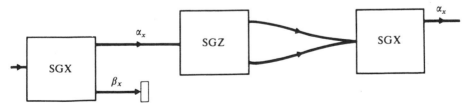

Figure 11.6 Spin-half particles with positive x component of spin pass through an SGZ apparatus and are then directed into an SGX without any record of their z component having been made. A further measurement of their x component invariably produces a positive result.

of classical physics. However, quantum mechanics is believed to be a universal theory, capable of describing macroscopic as well as microscopic objects, and it is when we try to extend the formalism of quantum mechanics to include the measuring apparatus that we get into difficulties. To understand this, we again consider the measurement of S_z described above, but now imagine that the passage of the particle is recorded by some detector and that this detector can itself be described by a wave function which can take one of three forms: χ_0 before any particle is detected, χ_+ if $S_z = \frac{1}{2}\hbar$, and χ_- if $S_z = -\frac{1}{2}\hbar$. We consider first the case where the initial spin state of the particle is represented by α_z; the total wave function of the whole system consisting of particle and measuring apparatus will then be ψ_0 where

$$\psi_0 = \alpha_z\chi_0 \tag{11.25}$$

After the particle has passed, χ_0 will be changed to χ_+ and the total wave function ψ_+ will be given by

$$\psi_+ = \alpha_z\chi_+ \tag{11.26}$$

Similarly if the initial state of the particle is β_z the final total wave function ψ_- will be given by

$$\psi_- = \beta_z\chi_- \tag{11.27}$$

We shall now consider what happens when the particle is not in an eigenstate of $\hat{S}_z$ before the measurement, but has a wave function whose spin part is α_x. The initial state of the whole system particle-plus-detector is then

$$\psi_0' = \alpha_x\chi_0 = \frac{1}{\sqrt{2}}(\alpha_z + \beta_z)\chi_0 \tag{11.28}$$

Now, if we are considering the whole system, we can assume that it is isolated from all external influences and the time-evolution of the wave function is therefore governed by the time-dependent Schrödinger equation with a time-independent Hamiltonian, the detailed nature of which we do not need to consider. It then follows that each term on the right-hand side of (11.28) evolves in time in exactly the same way as it does when the initial wave function is the appropriate eigenfunction of $\hat{S}_z$. Then, using (11.26), (11.27), and (11.28), we find that the total wave function after the measurement is ψ where

$$\psi = \frac{1}{\sqrt{2}}(\alpha_z\chi_+ + \beta_z\chi_-) \tag{11.29}$$

Most, if not all, of the difficulties associated with the quantum theory of measurement are associated with the above equation. First, the right-hand side of (11.29) implies that the detector is in none of the states χ_0, χ_+, or χ_-, but in a combination of the last two; thus, if the detector were to contain a pointer which could occupy one of three positions depending on which of these states it was in, then (11.29) would imply that the pointer was somehow delocalized between two positions. Although we have seen that quantum mechanics often implies that microscopic systems can be represented by wave functions of this type, it would be quite contrary to our everyday experience if classical macroscopic objects such as detectors were to be in such a state. Secondly, (11.29) appears to contradict the basic postulate concerning the reduction of the wave function (11.24) which implies that the wave function should have become *either* $\alpha_z\chi_+$ *or* $\beta_z\chi_-$ as a result of the measurement. The third difficulty arises as a result of an attempt to resolve the previous two: if the system of particle-plus-detector is to be treated as a single quantum system, then no reduction of the wave function should occur until a measurement is made on *this* system. We should therefore use some further apparatus to measure the state of the counter and after this the wave function will be reduced to $\alpha_z\chi_+$ or $\beta_z\chi_-$ depending on the result of this measurement. But this new measuring device could of course be considered as part of an even larger system which would then be isolated and whose wave function would again have a form similar to that of (11.21). There would therefore appear to be a potentially infinite chain of measurements and there would seem to be no point at which we can unambiguously state that the wave function has been reduced.

Schrödinger's Cat

The apparently paradoxical results discussed above are illustrated even more vividly when we discuss an example first suggested by Schrödinger in the early days of the subject. In this thought experiment, Schrödinger imagined the result of a quantum-mechanical measurement, such as obtaining the value $\frac{1}{2}\hbar$ from a measurement of S_z on a particle whose initial state was α_x, being used to fire a gun (or trigger some other lethal device) in the direction of an unfortunate cat which is consequently killed. If, however, the result is $-\frac{1}{2}\hbar$, the gun is not fired and the cat remains alive. The whole apparatus is enclosed in a box which is opened at a later time and the state of the cat (live or dead) examined, from which observation the value of S_z can be deduced. By analogy with the earlier discussion, the time-dependent Schrödinger equation predicts that the wave function of the box and all its contents will be given by (11.29) where χ_+ and χ_- now describe the whole apparatus (apart from the particle but including the cat) when the cat is alive and dead respectively. But this would imply that before the box is opened the state of the cat is neither alive nor dead! If we reject this obviously absurd conclusion we must ask at what point the wave function is reduced: is it when the particle enters the magnet (clearly not for reasons given above), when the gun fires, when the cat dies, or when?

We shall now outline several different approaches to the quantum theory of measurement which have been advocated in attempts to resolve the problems associated with the reduction of the wave function.

The Copenhagen Interpretation As stated earlier, what is generally accepted to be the conventional interpretation of the quantum theory of measurement is that developed by Niels Bohr and co-workers in Copenhagen and is consequently referred to as the Copenhagen interpretation. The essence of this point of view is that it is always incorrect to consider the quantum mechanical system as separate from the measuring apparatus. Thus a spin-half particle approaching an SGZ apparatus (including an appropriate counter) should be looked on as a completely different system from that consisting of the same (or a similar) particle approaching an SGX. The problem of reduction does not then arise as the choice of wave functions (α_z and β_z or α_x and β_x) is dictated by the details of the experimental arrangement. Moreover, if we remove the counter and bring the two paths together as discussed earlier, we change the experimental set-up, and we should therefore not be surprised that a corresponding change has taken place in the associated wave functions. This aspect of the Copenhagen interpretation receives powerful support from the theoretical analysis and experimental measurements performed on systems of correlated particles as discussed earlier in this chapter, where we saw that pairs of particles sometimes behave as a single system whose properties cannot be accounted for in terms of the separate properties of the individual particles.

Another powerful idea associated with the Copenhagen interpretation is that of *complementarity*. From this point of view some properties (e.g., position and linear momentum, x and z components of angular momentum, etc.) form complementary pairs, and it is an intrinsic property of nature that any attempt to define one of these variables precisely must lead to a complementary uncertainty in the other. For example, if a spin-half particle is in an eigenstate of S_x, to ask whether its spin is parallel or antiparallel to the z axis is clearly meaningless. In the same way, a proper understanding of quantum mechanics would imply that the question 'what is the position of a particle of known momentum?' is equally meaningless.

However, despite its great insights, the Copenhagen interpretation really does not address the problem of what constitutes a quantum measurement. Niels Bohr seems to have believed that the macroscopic apparatus is obviously classical with measurements corresponding to irreversible processes (see the sections on classical measurement and irreversibility below) while others in the Copenhagen school probably believed that measurement is an essentially human activity (see the section on subjective theories below).

Classical measurements The quantum measurement problem can be resolved quite straightforwardly if we postulate that classical mechanics, which provides an account of the behaviour of macroscopic objects, is not just a limiting case of quantum mechanics, but is a separate theory in which the results of measurements are always definite and unambiguous. Thus objects such as counters and cats are assumed to obey different laws from those followed by electrons or other microscopic particles; the reduction of the wave function associated with the latter then takes place as soon as these interact with the former, and, as these can never be treated quantum mechanically, the problems discussed above do not arise. The difficulties associated with this type of theory are, firstly, that it seems undesirably complicated and generally unreasonable to postulate two separate theories for the behaviour of material objects, depending on their size. The second difficulty concerns the distinction between microscopic and macroscopic objects as it is very difficult to draw a clear dividing line between these two classes. This is illustrated if we consider once again the properties of spin measurement in a Stern–Gerlach experiment. As we pointed out back in Chapter 5, this is normally performed on atoms (rather than obviously microscopic objects such as electrons) and passing these through a Stern–Gerlach magnet causes their possible paths to be separated to such an extent that, as far as their position and velocity are concerned, they are effectively classical objects. Moreover, there is no reason in principle why the experiment should not be carried out with the spins attached to much heavier bodies. Nevertheless, we showed above that quantum mechanics predicts that if the particles are deflected back so that we cannot tell which channel they passed through, then the spin part of their wave function is unchanged and it follows that those 'classical' objects must have been delocalized in a quantum sense while passing through the magnet. The fact that such delocalization is against all our everyday experience of macroscopic objects does indeed suggest that these do not obey quantum laws. However, if we ask whether there is any evidence to support this model rather than one of the alternatives discussed below, we would need to perform an interference experiment on a macroscopic body and fail to observe any interference when we should. It turns out in fact that this is extremely difficult; for example, unless great care is taken to ensure that the two possible paths through a Stern–Gerlach apparatus are kept constant to a high degree of accuracy (within about 10^{-6} m) the original spin function will not be reconstructed and the results of the experiment will be indistinguishable from one where the particles pass along one or other of the two possible paths without ever being delocalized. In fact the interference experiment illustrated in Fig. 11.5 has never been carried out with silver (or any other kind of) atoms, although the analogous experiment on polarized photons is relatively straightforward. It was pointed out by A. J. Leggett in the late 1970s that, as the magnetic flux in superconducting quantum interference devices (SQUIDS) is associated with the collective coherent motion of a macroscopic number of electrons, an observation of quantum interference involving this quantity would show that quantum mechanics is applicable to a macroscopic object in this context at least. The last ten years have seen the development of the theoretical implications of these ideas and some experiments. The latter have shown that magnetic flux can quantum mechanically tunnel through potential barriers, but definitive tests of its coherent delocalization are technically very difficult and have not yet been performed.

Irreversibility One of the key features of any practical measurement is that there is an irreversible change in the measuring apparatus at some point. Thus when the counter clicks or the grain in the photographic emulsion blackens, an irreversible change, accompanied by an increase in the entropy of the universe, occurs. If we postulate that quantum coherence breaks down whenever such a process occurs, the measurement problem can be at least partly resolved as we shall now show.

The first point to be made is that the *statistical* predictions for the results of experiments performed subsequent to an irreversible change are the same, whether or not we assume that reduction has taken place at this point. To see this, we consider the statistical results of a large number of identical SGZ measurements (including appropriate recording apparatus) performed on particles that are initially in the state α_x. Such a collection of identical experiments is known as an *ensemble* and the Schrödinger equation predicts that after the measurement the ensemble will be in a so-called 'pure state' described by a wave function of the form (11.21), while the reduction postulate implies that the ensemble is in a 'mixed state' where half the particles are known to be in each of the eigenstates of S_z, although it should be noted that which particle is in which state may be unknown.

We now consider how we can tell whether such an ensemble is in a pure or a mixed state. To do this we consider an operator $\hat{Q}$ which represents some physical operation on the whole system of particle-plus-measuring apparatus. If the system is in a pure state and the wave function has the form (11.29), the expectation value of $\hat{Q}$ will be given by $\langle \hat{Q} \rangle$ where

$$\langle \hat{Q} \rangle = \tfrac{1}{2} \int (\alpha_z^* \chi_+^* + \beta_z^* \chi_-^*) \hat{Q} (\alpha_z \chi_+ + \beta_z \chi_-) \, d\tau \qquad (11.30)$$

and the myriad of variables required to describe the state of the particle and measuring apparatus are all assumed to be included in the volume element $d\tau$. Multiplying out (11.30) we get

$$\langle \hat{Q} \rangle = \tfrac{1}{2} \int \alpha_z^* \chi_+^* \hat{Q} \alpha_z \chi_+ \, d\tau + \tfrac{1}{2} \int \beta_z^* \chi_-^* \hat{Q} \beta_z \chi_- \, d\tau + \text{Re}[\int \alpha_z^* \chi_+^* \hat{Q} \beta_z \chi_- \, d\tau] \qquad (11.31)$$

where Re means 'real part of'.

In order to calculate the detailed properties of mixed states, quantum-mechanical techniques using what is known as the *density matrix* must be employed. However, the only result required for our present purposes is the fairly obvious one that the expectation value in the mixed state is the mean of the expectation values calculated separately for the two sets of particles and apparatus whose wave functions are $\alpha_z \chi_+$ and $\beta_z \chi_-$ respectively. (Clearly, if there were not equal numbers of particles in each state, the mean would have to be weighted appropriately.) The expectation value in the mixed state is therefore given by $\langle \hat{Q} \rangle'$ where

$$\langle \hat{Q} \rangle' = \tfrac{1}{2}\int \alpha_z^* \chi_+^* \hat{Q} \alpha_z \chi_+ \, d\tau + \tfrac{1}{2}\int \beta_z^* \chi_-^* \hat{Q} \beta_z \chi_- \, d\tau \qquad (11.32)$$

Comparison of (11.31) and (11.32) shows that these two results would be identical if, and only if, the final term in (11.31) were to be zero. That is if

$$Q_{+-} \equiv \int \alpha_z^* \chi_+^* \hat{Q} \beta_z \chi_- \, d\tau = 0 \qquad (11.33)$$

Moreover, it can also be shown (although we shall not do so here) that a similar condition would ensure that the probability distributions relating to the possible outcomes of the measurement represented by $\hat{Q}$ would also be independent of whether the system was in a pure state or a mixed state. Thus, if terms of the form Q_{+-} were to vanish for all possible physical operators $\hat{Q}$, the mixed state and the pure state would be indistinguishable and we could choose whichever form was more convenient for the purposes of further calculation.

We now consider under what conditions two states, χ_+ and χ_-, are such that integrals of the form (11.33) always vanish. We note first that $|Q_{+-}|^2$ is proportional to the probability that a transition will take place between the two states χ_+ and χ_- under the influence of the operator $\hat{Q}$. If, therefore, this integral is always zero, it follows that no transition between the two states can ever take place. But this is just the same as saying that the process is irreversible. According to the second law of thermodynamics, an irreversible change leads to an increase in the entropy of the universe which can never be recovered. The only way we can restore the system, i.e. 'clear the counter', is to operate from outside the system in such a way that the entropy of some other part of the universe increases. Thus, as far as the *statistical* results of quantum mechanics are concerned, the reduction of the wave function becomes a matter of choice rather than necessity. It is possible to continue to represent the whole system of particle-plus-apparatus as being in a pure state describable by a wave function of the form (11.29), but once an irreversible change has occurred, it is also possible and generally much more useful to assume that reduction has taken place. We can then calculate the results of future measurements on the particle without having to include the very complex details of the state of the previous measuring apparatus. Extending these ideas from the particular example to the general behaviour of the material world, we see that reduction allows us to describe an isolated physical system without having to take into account the behaviour of all the other systems with which it has interacted irreversibly in the past—although if we could do so, the pure-state description would give identical results.

There are two main objections to the above argument as a complete resolution of the measurement problem. The first is that there is an arbitrariness about the distinction between a reversible and an irreversible change. The second law of thermodynamics is exactly true only in the 'thermodynamic limit' where the number of degrees of freedom of the system is infinite. For any finite system, however large, there is a correspondingly small, but finite, probability that an apparently irreversible change will actually be reversed, and if we wait long enough this will happen spontaneously. By 'long enough', we actually mean a time immensely longer than the age of the universe, so this objection has no practical consequences. The problem is to define objectively how large a system must be before we can say that the system is 'really' a mixture and the correlations can rigorously be ignored. One way past this has been suggested by the scientist Ilya Prigogine. He has pointed out that many-body macroscopic systems are very often chaotic on a molecular level, which makes the possibility of reversible behaviour vastly smaller even than that predicted by conventional statistical mechanics. Furthermore, he suggests that irreversibility should be explicitly built into the fundamental physical laws. The conventional laws of classical and quantum mechanics would then be approximations appropriate to the idealized case when a particle, or small number of particles, is isolated from the rest of the universe. The quantum measurement problem can be resolved in this way, but at the cost of a complete revolution in our traditional way of looking at the nature of the physical world, and until some direct experimental test of these ideas is devised and performed, it is unlikely that they will be generally accepted.

The second objection to irreversibility theory is that, even if the change is to a perfect mixture, this solves only half the problem, because it provides only a statistical result and says nothing about what happens to the individual objects. Even if the pure state is statistically indistinguishable from a mixture, what should it be a mixture of? There is nothing in the Schrödinger evolution that selects out the basic states $\alpha_z \chi_+$ and $\beta_z \chi_-$. Indeed the mixture could just as well be referred to the unreduced states when we would say that 100% of the particles are in the original state $2^{-1/2}(\alpha_z \chi_+ + \beta_z \chi_-)$. Intuitively, we say that it must be referred to the reduced basis, but this is just a prejudice formed from our experience of the classical world where detectors either click or do not click and cats are either dead or alive. We see now that, although we may not fully understand how the initial pure state is transformed into a mixture by a measurement, it is actually much more difficult to see why individual particles in the mixture are transformed into the new pure states. Even when a mixture is formed, Schrödinger time evolution says that each particle is in some sense in both states, but in reality we find that each particle is in one state or the other. The latter is of course just what the measurement postulate says, but underlying this is an assumption that the events whose statistics we are calculating are outside the Schrödinger formalism. This might be acceptable in the context of a hidden-variable model, but when we try to maintain that quantum mechanics is the fundamental, correct way of describing everything, we cannot avoid these problems without making additional assumptions.

Gravity Apart perhaps from quantum mechanics itself, the greatest theory of modern physics is Einstein's general theory of relativity. This has successfully united a model of space and time with gravity to provide a consistent picture of the large-scale structure of the universe and its evolution from the big bang. The predictions of general relativity have been thoroughly tested by a number of detailed astronomical measurements and have been confirmed each time. However, major problems arise when we try to unite gravity with quantum mechanics. This is because the relativistic view of space–time and its curvature in gravitational fields is essentially incompatible with the linearity of quantum theory. It is generally believed that significant problems could only arise in situations where the gravitational fields are typically as strong as those believed to exist in the neighbourhood of a black hole, i.e. immensely stronger than anything experienced on earth. However, the cosmologist Roger Penrose has speculated that it may be the very weak residual effects of quantum gravity that cause the collapse of the wave function when the system is in a macroscopically delocalized state. This will remain a speculation unless and until a full theory of quantum gravity that contains this result, along with others that can be tested experimentally, is developed.

Subjective theories One type of measurement theory which appears to overcome all the difficulties was suggested by E. P. Wigner. He pointed out that, in the last analysis, all we can know about the physical universe is the information that enters our mind through our senses and ends up in our consciousness. He therefore postulated that the reduction of the wave function occurs when the information enters a conscious human mind; thus the particle, the counter, and the cat are all in states described by expressions of the form (11.29), until a person opens the box and the information therein is transferred into a human mind; only at this point is the wave function reduced.

Although this theory can be made consistent with all the observed facts, it is generally considered unsatisfactory for a number of more or less obvious reasons. First, it relies heavily on the concept of the human mind or consciousness as something different in kind from the physical, material universe—including the brain. Certainly, many people, including some philosophers and scientists, believe this to be the case, but most would be unwilling to believe that the whole existence of the physical universe depends on this postulate: it should surely be possible to explain the natural world in objective terms consistent with it having an existence independent of our presence and interactions with it. Second, it pushes the whole problem into an inaccessible area; because if everything is 'in the mind' and this 'mind' is not a physical thing subject to investigation, then the whole of physics and science in general has no objective significance. Lastly, it is difficult to explain the fact that different 'minds' generally come to the same conclusions about the results of physical measurements (in both classical and quantum mechanics) unless we allow the existence of an objective physical universe.

Despite these objections and the fact that the reader will almost certainly find such subjective theories quite incredible, it should be remembered that they do provide an explanation for the quantum theory of measurement and that some scientists and philosophers believe that they provide the most satisfactory explanation available.

Many Worlds We complete our survey of measurement theories by considering one that is also based on the idea of a measurement being associated with an irreversible change in the universe. However in this version, first put forward by H. Everett, instead of the universe ending up in one of a number of possible states as a result of a quantum-mechanical measurement, it is suggested that all possible outcomes actually occur. To achieve this we have to postulate that a quantum-mechanical measurement causes the whole universe to divide or 'branch' into a number of separate non-interacting universes that can never be aware of each other's existence. Thus, according to this model, when we obtain an apparently random result from a quantum-mechanical experiment, all possible results have in fact occurred in a perfectly deterministic way; we observe a particular result in one branch of the divided universe, but other versions of ourselves are observing other results in other branches! Of course this branching is not confined to cases where experimental measurements are deliberately carried out, but must also be associated with every quantum event that occurs; thus the universe must be thought of as continually dividing into myriads of branches every second.

Many worlds theory therefore deals with most, if not all, of the problems associated with the quantum theory of measurement, but it does so in a particularly uneconomical manner. There is a fundamental postulate in science known as 'Occam's razor' which states that no theory should contain more postulates than are necessary to explain the observed facts. The idea of a near-infinite number of universes which can never interact with each other and whose existence can therefore never be verified seems to most scientists to be an extreme breach of this principle and certainly greater than that implied by the de Broglie–Bohm hidden-variable theory.

In recent years a combination of Many Worlds and subjectivism has been suggested. This proposes that there is no collapse; the wave function evolves according to the time-dependent Schrödinger equation and this is all there is in the physical universe. However, we conscious observers are incapable of seeing the world this way; it is in our nature that we can be aware of only one result of any measurement process so, although the others are still out there, we are unaware of them. One of the additional problems that this approach gives rise to is to explain how different conscious observers always see the same results, and indeed it has been suggested that this points to all our individual consciousnesses being linked to some 'universal consciousness', one of whose jobs is to ensure that this agreement occurs!

11.5 THE PROBLEM OF REALITY

It should be clear by now that one of the fundamental problems thrown up by quantum mechanics in general and the measurement problem in particular is the nature of reality—what it is that 'really exists' in the universe.

Throughout this book we have stressed that the wave function has no direct physical significance, but is a theoretical construct from which we can derive the relative probabilities of subsequent events. On the other hand, we showed earlier that, if we attribute properties to quantum systems in addition to those contained in the wave function, then (excluding the possibility of non-local hidden-variable theories) we obtain results that are in disagreement with both quantum mechanics and experiment. If the wave function is not physical and if there are no hidden variables, what can meaningfully be said to exist?

One answer to the above question is to return to subjectivism: the only things that we know exist are our sense impressions, so perhaps we should define these as the only reality. Or we could adopt the approach known as positivism in which, because it is impossible to verify the existence or non-existence of anything beyond our sense impressions, we describe such questions as meaningless. If, however, we define the aim of science as an attempt to understand and explain the properties of the natural world, on the assumption that in some sense it exists, we are required to consider the meaning of this statement.

One of the attractions of hidden-variable theories is that they contain a clear description of what really exists. Apart from this, perhaps the nearest we have come to an objective theory is the idea of the irreversible measurement as the fundamental reality. Perhaps it is the irreversible changes in the universe which have a real, objective existence: any genuinely reversible changes are incapable of observation, and statements about their existence or otherwise are consequently meaningless. Moreover, by describing such reversible changes as 'incapable of observation' we do not necessarily refer to the intervention of any human observer: such reversible 'events' simply have no effect on the ensuing behaviour of the universe. Quantum mechanics can therefore be thought of as a theoretical system whereby we can predict, as far as this is possible, the sequence of irreversible events in the universe; in this process we use quantities such as wave functions and talk about particles passing through slits without leaving any record, but it is only the irreversible changes which can be considered as having an objective existence. However, as we saw earlier, the irreversible theory also has problems and is still incomplete, because we cannot within quantum mechanics describe what it is that is changing or what it is changing into. Whichever way we look at these fundamental problems, we still come across unanswered questions.

It was never the purpose of this chapter to provide answers to the conceptual problems of quantum mechanics, but to demonstrate that such problems do exist and that real questions can be asked in this area. In earlier chapters we have seen many examples of the success of quantum mechanics in predicting and explaining experimental results over a wide range of physics, and most students of the subject will probably continue to be content to use it as a theoretical tool in their study and research into particular phenomena; there is, after all, no dispute concerning the correctness of quantum mechanics in predicting quantities such as energy levels, transition probabilities, scattering cross sections, etc. Others will prefer to set the subject aside and do something else. In any case it is to be hoped that all students will understand that there are still some real problems in the grey area where physics and philosophy meet. The influence of natural science on philosophy has been very considerable, particularly in recent years, and the further consideration of the fundamental problems of quantum mechanics may very well have far-reaching effects on our understanding of our natural environment and, eventually, of ourselves.

PROBLEMS

11.1 Show that an alternative proof of Eq. (11.10) can be obtained by defining the correlation coefficient $C(\phi)$ as the expectation value of the product $\hat{S}_{z1}\hat{S}_{\phi 2}$ divided by $\langle \hat{S}_{z1}\rangle\langle \hat{S}_{\phi 2}\rangle$ and using the fact that the wave function of a non-interacting two-particle system can be expressed as an antisymmetric linear combination of products of single-particle functions.

11.2 If the z component and total spin of a particle are known, then the 'real spin' vector, if it exists, must presumably lie on a cone whose symmetry axis is parallel to z. Using this hidden-variable theory, show that in the case of a spin-half particle with positive z component, the probability of a component in a direction at an angle ϕ to the z axis also being positive is equal to

$$
\begin{array}{cl}
1 & \text{if } \phi < \cot^{-1}(2^{1/2}) \\
1 - \pi^{-1}\cos^{-1}(2^{-1/2}\cot\phi) & \text{if } \cot^{-1}(2^{1/2}) \leqslant \phi < \cot^{-1}(-2^{1/2}) \\
0 & \text{if } \phi > \cot^{-1}(-2^{1/2})
\end{array}
$$

Show that this result agrees with quantum mechanics if ϕ equals 0, $\pi/2$, or π, but not otherwise. If this theory were applied to measurements on correlated pairs, would it be a local or a non-local hidden-variable theory?

11.3 Consider a hidden-variable theory based on the model described in Prob. 11.2 with the additional condition that the result of a subsequent measurement of a component in the direction defined by ϕ is to be determined by the sign of the component of the 'real spin' in a direction at an angle ϕ' to the z axis, where in the case of a spin-half particle

$$
\phi' = \cot^{-1}\{2^{1/2}\cos[2\pi\sin^2(\phi/2)]\}
$$

Show that this theory produces identical results to quantum mechanics, but does not preserve locality when applied to measurements on correlated pairs.

11.4 A variant of the Schrödinger's cat experiment, in which the cat is replaced by a human observer, was suggested by E. P. Wigner. This observer—known as 'Wigner's friend'—is not to be killed, however, but is to note and remember which of the possible quantum events occurs; subsequently the box is opened and Wigner makes a 'measurement' by asking his friend what happened. Discuss this procedure from the viewpoints of the different quantum theories of measurement, paying particular attention to the problem of the reduction of the wave function.

BIBLIOGRAPHY

References to the literature relating to particular points are made on occasions throughout the text. A guide to more general reading is given below.

Background

A more extensive treatment of the experimental evidence for the need for quantum mechanics than is given in Chapter 1 is contained in many textbooks on atomic physics, of which the following are good examples.

Eisberg, R. M., and Resnick, R., *Quantum Physics of Atoms, Molecules, Solids, Nuclei and Particles*, Wiley, New York, 1974.

Enge, H. A., Wehr, M. R., and Richards, J. A., *Introduction to Atomic Physics*, Addison-Wesley, Reading, Massachusetts, 1972.

Richtmeyer, F. K., Kennard, E. H., and Cooper, J. H., *Introduction to Modern Physics* (6th ed.), McGraw-Hill, New York, 1969.

A general familiarity with mathematical techniques, particularly calculus and elementary matrix algebra, up to a level typical of that of a first-year undergraduate physics course, is assumed; suitable textbooks covering this material are

Arfken, G., *Mathematical Methods for Physicists*, Academic, New York, 1985.

Boas, M. L., *Mathematical Methods in the Physical Sciences* (2nd ed.), Wiley, New York, 1984.

Complementary

Some of the textbooks on quantum mechanics which treat the subject at approximately the same level as the present volume are listed below with short comments.

Bohm, D., *Quantum Theory*, Prentice-Hall, New York, 1951. This thorough, discursive text is one of the few at this level to attempt a detailed discussion of the conceptual problems of the subject. It is of course rather out of date by now.

Dicke, R. H., and Wittke, J. R., *Introduction to Quantum Mechanics*, Addison-Wesley, Reading, Massachusetts, 1960. In many ways the approach of this book is rather similar to that adopted in the present work, but it relies on a more detailed understanding of formal classical mechanics and uses rather complex mathematical arguments at times.

Feynman, R. P., Leighton, R. B., and Sands, M., *The Feynman Lectures in Physics* (vol. III, *Quantum Mechanics*), Addison-Wesley, Reading, Massachusetts, 1965. This book contains many physical insights and is written in an attractive informal style. However, from the start it uses an abstract, matrix formulation which some students find difficult.

French, A. P., and Taylor, E. F., *An Introduction to Quantum Physics*, Nelson, Middlesex, 1978. This book describes the physical principles behind many quantum processes in considerable detail, and this discussion is illustrated by a large number of examples. However, detailed mathematical arguments are avoided, which somewhat limits its scope.

Mathews, P. T., *Introduction to Quantum Mechanics* (3rd ed.), McGraw-Hill, London, 1974. This book contains a rather condensed and mathematically based treatment of the basic ideas of quantum mechanics and their application to a number of physical problems.

Sillito, R. M., *Non-Relativistic Quantum Mechanics* (2nd ed.), University Press, Edinburgh, 1967. A good reliable text, if a little over detailed and mathematical in places, by the person who taught the subject to the present author.

Advances

The following is a small selection of the many books on quantum mechanics which treat the subject at a more advanced level.

Dirac, P. A. M., *Quantum Mechanics* (4th ed.), Oxford University Press, London, 1967. This classic text describes the formal, rigorous version of quantum mechanics that was first developed by its author.

Gillespie, D. T., *A Quantum-Mechanics Primer*, International Textbook Company, 1970. This short book contains a clear, simplified description of formal theory and is an excellent introduction to the book by Dirac (see above).

Landau, L., and Lifschitz, E. M., *Quantum Mechanics, Non-Relativistic Theory* (2nd ed.) (trans. J. B. Sykes and J. S. Bell), Addison-Wesley, Reading, Massachusetts, 1965. A long book which discusses a number of advanced topics in some detail.

Ziman, J. M., *Elements of Advanced Quantum Theory*, University Press, Cambridge, 1969. A good introduction to more advanced topics, particularly quantum field theory and the many-body problem.

The conceptual problems of quantum mechanics

The following constitute a selection of some publications on the topics discussed in Chapter 11. They all contain references to further published work.

Bellinfante, F. J., *Measurement and Time Reversal in Objective Quantum Theory*, Pergamon, Oxford, 1978. This book argues forcibly for quantum mechanics being an objective theory and for the importance of an 'indelible record' as a part of any measurement.

Clauser, J. F., and Shimony, A., 'Bell's theorem: experimental tests and implications', *Reports on Progress in Physics*, vol. 41, pp. 1881–1927, 1978. This review article contains a thorough discussion of Bell's theorem and carefully analyses the various experiments that have been performed to test it.

D'Espagnat, B., *Conceptual Foundations of Quantum Mechanics*, Benjamin, Massachusetts, 1976. This is probably the nearest there is to an authoritative reference work on this subject. It reviews and analyses the whole field, including non-separability, measurement theories, and the associated philosophical problems.

Hughes, R. I. G., *The Structure and Interpretation of Quantum Mechanics*, Harvard University Press, 1989. This book contains a thorough discussion of the whole field at a moderately advanced level.

Prigogine, I., *From Being to Becoming*, Freeman, San Francisco, 1980. This book develops Prigogine's ideas referred to in Chapter 11.

Rae, A. I. M., *Quantum Physics, Illusion or Reality*, University Press, Cambridge, 1986. This book discusses the conceptual problems of quantum mechanics at greater length, although at a somewhat less advanced level than is done in Chapter 11.

Applications

The applications of quantum mechanics discussed in this book are mainly, if not entirely, drawn from the fields of atomic, nuclear, particle, and solid-state physics. More detailed discussion of these subjects can be found in a number of textbooks of which the following are typical examples.

Corney, A., *Atomic and Laser Spectroscopy*, Oxford University Press, London, 1977.

Kuhn, H., *Atomic Spectra* (2nd ed.), Longman, Harlow. 1970.

Woodgate, G. K., *Elementary Atomic Structure* (2nd ed.), Oxford University Press, London, 1980.

Burcham, W. E., *Nuclear Physics: An Introduction* (2nd ed.), Longman, Harlow, 1973.

Cohen, B. L., *Concepts of Nuclear Physics*, McGraw-Hill, New York, 1971.

Perkins, D. H., *Introduction to High-Energy Physics*, Addison-Wesley, Reading, Massachusetts, 1972.

Hook, J. R., and Hall, H. E., *Solid State Physics*, Wiley, 1991.

Kittel, C., *Introduction to Solid State Physics* (5th ed.), Wiley, New York, 1976.

HINTS TO SOLUTION OF PROBLEMS

Bracketed numbers, e.g. (1.1), refer to equations in the main text.

Chapter 1

1.1 Use Eq. (1.1): answer $h = 6.4 \times 10^{-34}$ J s, $\phi = 1.9$ eV.

1.2 Calculate energy arriving per second at a potassium atom ($4\pi \times 10^{-19}$ W). Hence time for 2.1 eV to arrive is 0.27 s. There are about 10^{16} potassium atoms in 10^{-3} m^2 so average emission rate is about 4×10^{16} electrons s^{-1}. This would not be affected by quantum considerations.

1.3 Use (1.7): answers (i) 2.213×10^{-12} m; (ii) 3.426×10^{-12} m; (iii) 4.639×10^{-12} m.

1.4 Use (1.10) to show that electron wavelength is 1.23×10^{-10} m. Use (1.11) to calculate Bragg angles; answer $\sin^{-1}(0.123n)$. For neutrons of the same energy, wavelength is too large for diffraction to occur.

1.5 See discussion in Sec. 4.5.

Chapter 2

2.1 (i) Use (2.26): answers 1.05 eV; 4.19 eV; 9.43 eV. (ii) Use answers to (i) and (1.2): answer 1.47×10^{-7} m. (iii) Use answers to (i) and (1.2): answers 1.47×10^{-7} m; 2.35×10^{-7} m; 3.92×10^{-7} m.

2.2 Using (2.27) with n and m both even,

$$\int_{-\infty}^{\infty} u_n u_m \, dx = a^{-1} \int_{-a}^{a} \cos(n\pi x/2a) \cos(m\pi x/2a) \, dx$$
$$= \tfrac{1}{2} a^{-1} \int_{-a}^{a} (\cos[(n+m)\pi x/a] + \cos[(n-m)\pi x/a]) \, dx$$
$$= 0 \text{ by symmetry if } n \neq m$$

Similar results for n and/or m odd are proved similarly.

254

2.3 At $x = 0$, boundary condition requires $u = 0$. At $x = a$, boundary condition requires u and du/dx to be continuous. Proceed as for finite square well, cf. second of (2.42).

2.4 Relevant equation is the second of (2.44) with $k_0 a = 3$. Iteration with $n_2 = 1$ leads to $ka = 2.279$ so that $E/V_0 = 0.865$. No real solutions exist for $n_2 > 1$ so this is the only bound state.

2.5 Putting $x = -y$ and using $V(x) = V(-y) = V(y)$ we find that the Schrödinger equation for $u(-y)$ is the same as that for $u(y)$. Hence solutions of the two equations must be the same apart from a multiplicative constant. But if $u(x) = Au(-x)$, then $u(x) = A^2 u(x)$ and $A = \pm 1$.

2.6
$$u = A \cos kx \qquad -a \leqslant x \leqslant b$$
$$u = C \exp(-\kappa x) + D \exp(\kappa x) \qquad a \leqslant x \leqslant b$$
$$u = C \exp(\kappa x) + D \exp(-\kappa x) \qquad -b \leqslant x \leqslant -a$$
$$u = B \cos kx \qquad |x| > b$$

Apply boundary conditions at $x = a$ and $x = b$. Unless $D = 0$, condition $k(b - a) \gg 1$ implies that $A \ll B$. If $D = 0$, boundary conditions are as for a finite well and we get

$$B/A = (\cos ka/\cos kb) \exp[-\kappa(b - a)]$$

2.7 Transition must be between neighbouring energy states, where the energy difference is $\hbar\omega_c$. Hence $\omega_c = 4.71 \times 10^{14}$ rad s^{-1}, zero-point energy $= \frac{1}{2}\hbar\omega_c = 2.48 \times 10^{-20}$ J; $\hbar\omega_c \simeq k_B T$ when $T = 3600$ K so ground state is most probable at $T = 450$ K.

2.8 Use (2.19) and (2.59) along with definition of x and standard expressions for integrals. Answers as in (2.60).

Chapter 3

3.1 Energy levels as (3.14) with $n_3 = 0$; wave functions as (3.15) with $n_3 = 0$ and $c = 1$. Degeneracy discussion identical to that following (3.16) and (3.17).

3.2 Position probability distribution has full cubic symmetry because it equals the average of the squared wave functions of all the degenerate states with the same value of $n_1^2 + n_2^2 + n_3^2$.

3.3 Start from plane-polar expression for

$$\nabla^2 = \frac{1}{r} \frac{\partial}{\partial r}\left(r \frac{\partial}{\partial r}\right) + \frac{1}{r^2} \frac{\partial^2}{\partial \phi^2}$$

Separate the variables as in the three-dimensional case and get an equation in ϕ that is the same as (3.29). The position probability distribution is circularly symmetric.

3.4 Equation follows from separation of variables as in Prob. 3.3. Recurrence relation follows from series solution. Using boundary condition at $r = a$ we get $E = 2.892/\mu a^2$.

3.5 If $l = 0$, Eq. (3.47) for $\chi(r)$ is identical to (2.16) for $u(x)$. Hence possible solutions are the same as for 1-D finite square well *except that* $\chi(r)$ must $= 0$ at $r = 0$. Thus only solutions with $A = 0$ are allowed and these are determined by $k \cot ka = -\kappa$. Hence $k > \pi/2a$ and condition follows. If $l \neq 0$, V would have to be more negative for the state to be bound. There is no such condition in a cubic well.

3.6 Binding energy is $V_0 - E$. With $x = ka$ the energy equation is as given. Hence $k = 1.82/a$; $E = 34.43$ MeV; $V_0 = 36.66$ MeV. As the ground state is only just bound, no other bound states can exist.

3.7 Use expressions in (3.46).

3.8 Probability that r be between r and $r + dr$ is $4\pi r^2 u_{100}^2$. Use (3.70). Answers are (i) a_0, (ii) $3a_0/2$.

Chapter 4

4.1 Clearly $[\hat{Q}^n, \hat{Q}^m] = 0$ where n and m are any integers. K.E. $= (\hat{P}_x^2 + \hat{P}_y^2 + \hat{P}_z^2)/2m$. Hence $[\hat{P}_x, \hat{V}] = 0$ only if $\partial V/\partial x = 0$ everywhere.

4.2 Equation can be written as

$$\psi = 5^{-1/2}(u_1 + 2u_2)$$

Possible energies are E_1 with probability $1/5$ and E_2 with probability $4/5$ (cf. 2.26). After measurement $\psi = u_1$ or u_2 and we are certain to repeat the same result for E.

4.3 Form the Fourier transform of u and square its modulus. Normalize so that

$$\int_{-\infty}^{\infty} P(k)\,dk = 1$$

4.4 Repeat 4.3 for a highly excited state where (e.g.) $\psi = a^{-1/2}\cos(n\pi x/2a)$ (n large and odd). Denominator is now $(n^2\pi^2/4a^2 - k^2)$. The result follows and is the same as in the classical case.

4.5 Original $\psi = a^{-1/2}\cos(\pi x/2a)$ (zero if $|x| > a$). New $\psi = (2a)^{-1/2}\cos(\pi x/4a)$ (zero if $|a'| > 2a$) in ground state case. Use (4.30) to show that the probability of finding system in new ground state is A^2 where

$$A = \int_{-a}^{a} (2a^2)^{-1/2}\cos(\pi x/2a)\cos(\pi x/4a)\,dx = 0.849$$

Hence $A^2 = 0.721$ and other probabilities are similarly calculated as 0 and 0.259.

4.6 Use (4.33) and Sec. 2.66. $\langle x \rangle = 0$; $\langle x^2 \rangle = \hbar/2m\omega_c$; $\langle p \rangle = 0$; $\langle p^2 \rangle = \frac{1}{2}\hbar m\omega_c$; $\langle E \rangle = \frac{1}{2}\hbar\omega_c$; $\langle x^2 \rangle^{1/2}\langle p^2 \rangle^{1/2} = \frac{1}{2}$(cf. 4.42).

4.7 Use (3.70). $\langle r \rangle = 3a_0/2$; $\langle r^2 \rangle = 6a_0^2$; $\langle V \rangle = -e^2/2\pi\varepsilon_0 a_0$; $\langle T \rangle = e^2/4\pi\varepsilon_0 a_0$; $\langle H \rangle = e^2/4\pi\varepsilon_0 a_0$ in agreement with (3.65).

4.8 Use (4.11) and make substitution $y = -x$. Eigenvalues follow from $\hat{P}^2\phi = p^2\phi = \phi$. Eigenfunctions are then obvious.

4.9 If $\hat{H}$ is centrosymmetric it is unaffected by $\hat{P}$ so $[\hat{P}, \hat{H}] = 0$. In the non-degenerate case, $\hat{H}$ and $\hat{P}$ have a common set of eigenfunctions. In the degenerate case eigenfunctions of definite parity can be chosen, but are not obligatory.

4.10 Three-dimensional case similar to one-dimensional case. Spherical harmonics have parity $(-1)^l$.

4.11 From (4.55) we can determine the component of $\mathbf{P}$ in the direction of $\Delta\mathbf{P}$ which is at about $45°$ to the direction of the incident X-rays. Using (3.2) and (4.55) the magnitude of this component is 8.5×10^{-24} kg m s^{-1}.

Chapter 5

5.1 Classically $E = L^2/2ma^2$; answer follows from (5.18).

5.2 $E = L_z^2/2ma^2 = n^2\hbar^2/2ma^2$ where n is an integer; other components of momentum are unknown from (5.5)–(5.7).

5.3 Use (4.5), (4.10), (5.3) and (5.4); answers $L_z = 0$, $L^2 = 2\hbar^2$ (i.e. $l = 1$, $m = 0$). Then use (5.30); answers $(x \pm iy)\exp[-a(x^2 + y^2 + z^2)]$.

5.4 Use (3.46), (5.16) and (5.17).

5.5 Separation is $z = Fdl/2T$ where F is magnetic force, d is distance travelled through magnet, l is distance travelled beyond magnet and T is kinetic energy. Answer 2.67 mm.

5.6 Use (5.16); $L_z = 0$.

5.7 Use (5.30) and the fact that $\hat{L}_x$ and $\hat{L}_y$ are Hermitian.

5.8 Put normalized $\phi_{m+1} = A\hat{L}_+\phi_m$. Use the result of 5.7 to show that

$$\int |\phi_{m+1}|^2 \, d\tau = A^2 \int \phi_m^* \hat{L}_- \hat{L}_+ \phi_m \, d\tau$$

The result follows by expressing $\hat{L}_-\hat{L}_+$ in terms of $\hat{L}^2$ and $\hat{L}_z^2$.

5.9 $\hat{L}_x = \frac{1}{2}(\hat{L}_+ + \hat{L}_-)$; $\hat{L}_y = \frac{1}{2}(\hat{L}_+ - \hat{L}_-)$. Hence $\langle \hat{L}_x \rangle = \langle \hat{L}_y \rangle = 0$.
$\langle \hat{L}_x^2 \rangle = \frac{1}{2}\langle \hat{L}_+\hat{L}_- + \hat{L}_-\hat{L}_+ \rangle = [l(l+1) - m^2]\hbar^2 = \langle \hat{L}_y^2 \rangle = \langle \hat{L}_x^2 \rangle^{1/2} \langle \hat{L}_y^2 \rangle^{1/2}$.
From uncertainty principle $\langle \hat{L}_x^2 \rangle^{1/2} \langle \hat{L}_y^2 \rangle^{1/2} \geqslant \frac{1}{2}\hbar \langle \hat{L}_z \rangle = \frac{1}{2}m\hbar^2$.

5.10 Unit vector in x direction is $\mathbf{r}_0 \sin\theta\cos\phi + \boldsymbol{\theta}\cos\theta\cos\phi - \boldsymbol{\phi}\sin\phi$. Use with (5.15) to prove result.

5.11 Apply $\hat{L}_+$ and $\hat{L}_-$ from (5.10) to get unnormalized expressions, then normalize using (5.8). Results the same as in (3.46) apart from some sign changes.

Chapter 6

6.1 Use standard rules for matrix multiplication. The matrix for $[H]$ is diagonal with energy levels along the diagonal.

6.2 Use (6.15) and (6.16) and Table 6.1. Answers $\langle \hat{S}_x \rangle = \langle \hat{S}_y \rangle = 0$; $\langle \hat{S}_x^2 \rangle = \langle \hat{S}_y^2 \rangle = \frac{1}{16}\hbar^4$.
From the uncertainty principle $\langle \hat{S}_x^2 \rangle \langle \hat{S}_y^2 \rangle \geqslant \frac{1}{4}\langle [S_x, S_y] \rangle^2 = \frac{1}{4}\hbar^2 \langle S_z^2 \rangle = \hbar^4/16$.

6.3 $[S_+] = \hbar\begin{bmatrix} 0 & 1 \\ 0 & 0 \end{bmatrix}$ $[S_-] = \hbar\begin{bmatrix} 0 & 0 \\ 1 & 0 \end{bmatrix}$

Apply to $[S_z]$ eigenvectors to confirm properties.

6.4 Generalize the argument leading to (6.26) and (6.27). The probability of $\frac{1}{2}\hbar$ is $\cos^2\psi$ where ψ is the angle between the x axis and the measurement direction—i.e. $\cos\psi = \cos\theta\sin\phi$.

6.5 This question is discussed in Chap. 11, Sec. 11.3.

6.6 Apply the usual rules of matrix algebra. Eigenvalues are $\hbar$, 0 and $-\hbar$.

6.7 Eigenvectors of matrices in 6.6 are:

for L_x $\frac{1}{2}(1, -2^{1/2}, 1), 2^{-1/2}(1, 0, -1), \frac{1}{2}(1, 2^{1/2}, 1)$;
for L_y $\frac{1}{2}(1, 2^{1/2}i, -1), 2^{-1/2}(1, 0, 1), \frac{1}{2}(1, -2^{1/2}i, -1)$;
for L_z $(1, 0, 0), (0, 1, 0), (0, 0, 1)$.
The probabilities are the squares of the products of corresponding eigenvectors = $1/4$ for $\hbar$ and $-\hbar$ and $1/2$ for zero.

6.8 Follow the procedure described in the text. Answer:

$l_1 = 3$		$l_2 = 1$	
(m_1, m_2)		m	l
(3, 1)		4	4
(3,0) (2,1)		3	4, 3
(3, −1) (2, 0) (1, 1)		2	4, 3, 2
(2, −1) (1, 0) (0, 1)		1	4, 3, 2
(1, −1) (0, 0) ((−1, 1)		0	4, 3, 2
(−2, 1) (−1, 0) (0, −1)		−1	4, 3, 2
(−3, 1) (−2, 0) (−1, −1)		−2	4, 3, 2
(−3, 0) (−2, −1)		−3	4, 3
(−3, −1)		−4	4

$l_1 = 1$		$l_2 = 3/2$	
(m_1, m_2)		m	l
(1, 3/2)		5/2	5/2
(1, 1/2) (0, 3/2)		3/2	5/2, 3/2
(0, 1/2) (1, −1/2) (−1, 3/2)		1/2	5/2, 3/2, 1/2
(0, −1/2) (−1, 1/2) (1, −3/2)		−1/2	5/2, 3/2, 1/2
(−1, −1/2) (0, −3/2)		−3/2	5/2, 3/2
(−1, −3/2)		−5/2	5/2

6.9 Spin-orbit coupling as in Fig. 6.2. Weak field splits the $j = 5/2$ level into six components and $j = 3/2$ level into four. In a strong field there are seven components corresponding to the different possible values of $m_l + 2m_s$.

6.10 From Table 6.2, spin-orbit coupling splits level into three. These further divide into a septuplet, quintet and a triplet in a weak field. In a strong field we get a total of nine components.

Chapter 7

7.1 Use (7.12) with (2.27). Remember integrals are between $-a$ and a. Answers: $8V_0/8\pi$; $32V_0/15\pi$; $72V_0/35\pi$.

7.2 As in 7.1. Answers $V_0[b/a + \pi^{-1} \sin(\pi b/a)]$; 0; $V_0[b/a + (3\pi)^{-1} \sin(3\pi b/a)]$.

7.3 The field outside a spherical shell is the same as that outside a point charge so perturbation is zero there. Inside the shell, the potential is uniform and equal to $e/4\pi\varepsilon_0\delta$. $\hat{H}'$ is the difference between this and the point charge potential. Assume unperturbed wave function constant inside the nucleus. Ratio of perturbation to unperturbed energy is $4\delta^2/3a_0^2 = 4.76 \times 10^{-10}$.

7.4 All matrix elements $H'_{nn'}$ are zero unless $n = 1$ and $n' = -1$ when $H'_{nn'} = V_0/2$. Hence all first-order changes are zero except to the degenerate pair $n = \pm 1$ where the changes are $V_0/2$. Also $H'_{0n} = 0$ unless $n = 2$ when $H'_{0n} = V_0/2$; use (7.20).

7.5 Perturbation (7.41) has odd parity and the square of any function of definite parity is even, hence first-order integrals (7.13) vanish if starting functions have even parity. The Stark effect is observable in hydrogen because of mixing of two degenerate levels of different parity.

7.6 Use (6.58) choosing the appropriate unperturbed eigenfunctions to have quantum numbers n, j, l and s. $\delta E = 3\hbar^2 \langle f \rangle = 2\pi\hbar c\delta\lambda/\lambda^2$, so $\langle f \rangle = 7.18 \times 10^{-22}$ C kg^{-1} m^{-2}.

7.7 Get $R_{21} = (24a_0)^{-1/2}(r/a_0) \exp(-r/2a)$ from (3.70) and (3.46); and $f = e^2/(16\pi\varepsilon_0 m^2 c^2 r^3)$ from (6.56) and the Coulomb potential. The result follows.

7.8 Use (6.61) remembering that $(\hat{J}^2 - \hat{L}^2 - \hat{S}^2) = 2\hat{\mathbf{L}} \cdot \hat{\mathbf{S}}$.

7.9 This has the same form as the exact expression and therefore leads to $E_0 = \frac{1}{2}\hbar\omega_c$ and $u_0 = (m\omega_c/\pi\hbar)^{1/4} \exp(-m\omega_c x^2/2\hbar)$.

Chapter 8

8.1 Use (8.4) to get

$$|\psi(x, t)|^2 = \frac{1}{2a} [\cos^2(\pi x/2a) + \sin^2(\pi x/a) + 2 \cos(\pi x/2a) \sin(\pi x/a) \cos(3\pi^2\hbar t/8ma^2)]$$

8.2 Use (8.5).

8.3 Form $\langle \hat{S}_x \rangle$, $\langle \hat{S}_y \rangle$ and $\langle \hat{S}_z \rangle$ using (8.11).

8.4 Use (8.17) with (3.70). Answers $512/729$; $1/4$; 0.

8.5 Use (8.17) with results from Chapter 2.
Answers $2^{5/4}(1 + \sqrt{2})^{-1}$; 0; $2^{1/4}/(\sqrt{2} - 1)(1 + \sqrt{2})^{-3}$.

8.6 Calculate the speed from the wavelength using (1.10) and show that the flight time is correct for a relative precession of $360°$ to have occurred.

8.7 Use (8.33) and (2.27) and symmetry to show that integrals vanish when the condition is not fulfilled.

8.8 $\hat{H}''$ consists only of terms proportional to $\hat{L}_+$, $\hat{L}_-$ and $\hat{L}_z$. When operated on an angular-momentum eigenstate $\hat{H}''$ therefore changes it to a linear combination of angular-momentum eigenstates whose quantum numbers are related to the original ones by the expressions given. All other matrix elements therefore vanish by orthogonality. (i) $\Delta l = 0$, $\Delta m = 0$; (ii) $\Delta l = 0$; $\Delta m = \pm 1$.

8.9 As $\hat{H}''$ is proportional to y results follow directly from orthogonality.

8.10 Equations follow if the time-dependent Schrödinger equation is multiplied by u_1^* and integrated and then by u_2^* and integrated. Expressions for a and b follow if the cosine is expressed as a sum of exponentials.

8.11 Lowest eigenfunction is symmetric, next lowest is antisymmetric. In both cases the wave function is small within the centre barrier and there are no modes in either of the wells. From a given starting condition, the wave function will oscillate from side to side with a frequency corresponding to the energy difference between the states. Response to perturbation is very like that of the ammonia maser.

Chapter 9

9.1 Assume the wave function to be of the form:

$$A e^{ikx} + B e^{-ikx}, x < 0; \qquad C e^{ik_2 x} + D e^{-ik_2 x}, 0 < x < a; \qquad E e^{ikx}, x > a$$

Apply boundary conditions in the usual way (cf. Chap. 2) and eliminate C, D and E. Transmission probability is $1 - P$.

9.2 Rearranging,

$$P = U_0^2 \sin^2 k_2 a / [U_0^2 \sin^2 k_2 a + 4E(E - U_0)] = 0 \qquad \text{if } k_2 a = n\pi$$

Condition for maximum follows from equating dP/dE to zero, remembering that k_2 is a function of E. Minimum scattering case is very similar to Ramsauer–Townsend and maximum condition is similar to that applying to resonant scattering.

9.3 The same as (9.1) and (9.2) with k_2 defined as $2m[(U_0 - E)/\hbar^2]^{1/2}$ so that $\sin k_2 a$ is replaced by $\sinh k_2 a$ in all expressions.

9.4 Follow the same argument as for equations (9.28) to (9.34), but in one dimension. Both methods lead to $P' = U_0^2 \sin^2 ka/4E^2$.

9.5 Use (9.34) in Cartesian coordinates.

9.6 Use (9.36). Answer $\pi^2 m^2 A^2 / \hbar^4 K^2$.

9.7 If k and k' are both small, (9.72) becomes $\cot \delta_0 = 3/[k(k'^2 - k^2)a^3]$. The result follows using (9.74) and remembering that (9.39) is a differential cross section.

9.8 Follows directly from the fact that the equivalent of (9.70) has 'sin' replaced by 'sinh' and 'cos' by 'cosh'. If U_0 is very large, $\tanh k'a = 1$, $k' \gg k$ and (9.72) becomes equivalent to (9.67).

9.9 Use (9.77) with $E = 0$. Calculating σ from (9.77) with $E_b = -70$ keV gives 74×10^{-28} m². $74/4 + 3 \times 2.4/4 = 20.3$.

Chapter 10

10.1 Apply commutation relations to (10.2) and (10.3) remembering that no external forces implies

$$\sum_i \partial V/\partial x_i = 0 \qquad \text{etc.}$$

10.2 Use (10.8) and (10.9) along with (2.26) and (2.27)

$$E_n = P^2/2(m_1 + m_2) + \hbar^2\pi^2 n^2/8\mu a^2$$

$u_n = a^{-1/2} \cos[n\pi(x_1 - x_2)/2a] \exp[iPx/2\hbar]$ if n is odd where x is the centre of mass (cf. 10.5). Solutions with n even are the same with 'cos' replaced by 'sin'.

10.3 Answers as in Prob. 10.2 with $m_1 = m_2$, but in the boson case only allowed solutions are those with n odd while in the fermion case only even-n solutions are allowed.

10.4 The first part follows from probabilistic interpretation of ψ remembering that all particles are equivalent. The second part follows from the first and (10.22) or (10.24) with orthogonality.

10.5 Ground state has eigenfunction $2^{-1/2}u_1(x_1)u_1(x_2)[\alpha(1)\beta(2) - \alpha(2)\beta(1)]$ where $u_1(x) = a^{-1/2}\cos(\pi x/2a)$. Four-fold degenerate first excited state has eigenfunctions similar to last three of (10.28) with $u_2(x) = a^{-1/2}\sin(\pi x/a)$. Energies are $\pi^2\hbar^2/4ma^2$ and $5\pi^2\hbar^2/8ma^2$ respectively. Second part follows directly putting $x_1 = x_2$.

10.6 In the absence of interactions we can assign two opposite-spin electrons to each single-particle state with n between 1 and $N/2$. Hence

$$E = 2(\pi^2\hbar^2/2mN^2a^2) \sum_{n=1,N/2} n^2 \simeq (\pi^2\hbar^2/24ma^2)N$$

10.7 Substitute into the expression given in Prob. 10.6: answers (i) 0.503 eV; (ii) 9.1×10^{-5} eV. In ^{4}He gas all atoms are in ground state and energy is very small. Temperature corresponding to (ii) is 1.1 K. Note that the superfluid transition temperature in ^{4}He $\simeq 2$ K.

10.8 For s-wave scattering by identical particles (9.77) holds. Hence $E_b = 89$ keV.

Chapter 11

11.1 Take $\psi = 2^{-1/2}[\alpha(1)\beta(2) - \alpha(2)\beta(1)]$ and $\hat{S}_{\phi2} = S_{\phi2}\cos\phi + \hat{S}_{x2}\sin\phi$. Use orthonormality.

11.2 In the case of spin-$\frac{1}{2}$, the semi-angle of cone is $\tan^{-1}(2^{1/2})$. Hence if the spin axis lies along this cone it cannot have a component in the negative ϕ direction of $\phi < 90°$ — cone angle, i.e. $\phi < \cot^{-1}(2^{1/2})$. Similarly it has no positive ϕ component if $\phi > \cot^{-1}(-2^{1/2})$. In the intermediate case, the expression corresponds to the fraction of the cone surface pointing in the positive ϕ direction.

11.3 Substitute expression into that given in Prob. 11.2 which then reduces to $\cos\phi$ as predicted by quantum theory. The theory is deterministic and agrees with quantum theory and so cannot preserve locality by Bell's theorem.

INDEX